ECLIPSE

VOYAGE TO DARKNESS AND LIGHT

The eclipse enters totality with the appearance of the Sun's corona, on the Atlantic Ocean 350 kilometers southeast of Halifax, Nova Scotia. Photo by Roy Bishop.

ECLIPSE

VOYAGE TO DARKNESS AND LIGHT

DAVID H. LEVY

iBooks

new york

www.ibooksinc.com

HABENT SUA FATA LIBELLI

Dedication

For Mom and Dad, who kindled my passion for eclipses as a
child. Mom dashed across town to find a clear spot so that I
could see my first eclipse in 1959; and my parents went to great
lengths to make sure I saw my first total eclipse in 1963.
For Wendee, who rekindled my passion for eclipses and who
made the 1998 and 1999 eclipses cherished memories.
And for Nannette, Mark, Summer, and all the eclipses in their future.

An Original Publication of ibooks, inc.

Copyright © 2000 ibooks, inc.
Text copyright © 2000 David H. Levy

Frontispiece: The August 11, 1999 solar eclipse enters totality,
as seen from the Atlantic Ocean, 217 miles southeast
of Halifax, Nova Scotia.
Photo: Roy Bishop

An ibooks, inc. Book

iBooks
Manhanset House,
Shelter Island Hts, NY 11965-0342
iBooks are an imprint of J. Boylston & Company, Publishers

ISBN 978-1-59687-701-6

First ibooks, inc. printing November 2000
September 2018

Cover design by Mike Rivilis
Interior design by Michael Mendelsohn at MM Design 2000, Inc.

ACKNOWLEDGMENTS

I n writing a book about a total eclipse of the Sun, I must first thank the Sun, the Moon, and the Earth for arranging themselves so precisely as to allow this event to take place at all. Unheard of and unseen anywhere else in the solar system, total solar eclipses, where Moon covers Sun and only the Sun, are unique to Earth and are quite probably rare throughout the Universe. We are extraordinarily lucky.

On human terms, I begin by thanking my wife Wendee, who revitalized my interest in eclipses. After the almost seven minutes of totality in Mexico in 1991, I thought I had seen every possible thing during a total eclipse, and I didn't care if I never saw another. But then came the annular eclipse of 1994. I observed with Clyde and Patsy Tombaugh in Las Cruces, New Mexico, and Wendee, then teaching physical education a few miles away, developed a program during which her young charges could watch the eclipse in safety. Afterwards, when she asked me what the difference between an annular eclipse and a total eclipse, I began to explain how, in the eclipse she saw, the Moon did not appear large enough to cover the Sun, and that in a total eclipse. . . . Suddenly I realized that the only way to answer her question was that we both had to see a total eclipse. We saw the eclipse of February 26, 1998, which to me was more amazing than any other, and we were hooked. Wendee's family also had a great deal to contribute to the success of the 1999 voyage and the book that followed it. Her parents, Leonard and Annette, her sisters Gail and Joan-ellen, and her brother Sandy, all enjoyed this, their first eclipse—except for Annette, who saw the 1925 eclipse from New York City.

On the *Regal Empress*, Ann Burgess of North Star Cruises did a superb job organizing the cruise and encouraging Regal Cruises to sail into the path as expertly as they did. Joe Rao and Sam Storch, two of the lecturers, provided important background. Captain Peter Schaab and his crew did a magnificent job building the experience that made this book a possibility. On the editorial side, my editor, Howard Zimmerman, and publisher Byron Preiss, came up with the idea for this book and helped it through to completion with enthusiasm and finesse. I also thank Art Boehm for his editorial suggestions.

Finally, thanks to Roy and Gertrude Bishop, and Leo Enright and Denise Sabatini, and Patsy Tombaugh. They shared this voyage at the same time they cemented our special bond of friendship.

CONTENTS

CONTENTS

PREFACE

F ive billion years ago, a huge cloud of hydrogen gas swirled around in space. The cloud was thickest at its center, so dense that the gas glowed from the heat of its own gravity. At the outskirts of the cloud were other concentrations of gas and dust. As the cloud spun faster, its center grew hotter, until, at one single moment in time, it ignited in a burst of nuclear fusion to begin the life of our Sun.

With that ignition, the remaining gas and dust in the cloud's outlying regions congealed to build at least nine planetary worlds, several dozen smaller worlds, and many comets and asteroids. At least two of these worlds were about the same size, one-fourth that of Earth. One moved around the Sun in a lazy elliptical path. It still does; we call it Mars. The other world flew round the Sun in an elongated orbit, alternately moving out into the fringes of the solar system and then tearing through the realm of the inner planets. Many times in that primordial history, this world came so close to Earth that its bulk would fill half the sky as it raced by, causing massive earthquakes and tidal disruptions before moving away. Finally, on one frightening day some four and a half billion years ago, this world came too close, sideswiping, moving away briefly, and then slamming into Earth.

The resulting explosion was so vast that it melted the entire crust of Earth. The other world broke apart, its pieces spinning out of control to collide with each other, and with Earth, again and again as a ring of debris, from both the Earth and the doomed world, formed high above Earth. Over the next year, the ring's particles collided with one another, accreting to form at least one, and possibly two, new moons. One of them, too close, broke apart again, its pieces colliding with Earth in still more devastation. The other, farther away, formed the Moon we know and love.

At its creation the Moon was so close to Earth, a mere 10,000 miles away, that its immense tidal pull caused massive earthquakes and volcanic eruptions on a daily basis. As time passed, the Moon crept slowly farther from Earth, and still does, at the rate of about one and a half inches per year.

As it moved outward, the Moon's gravity affected the tides of Earth less and less. As it took up less space in the sky, the Moon seemed to shrink, gradually becoming the same apparent size as the Sun, which is much bigger but also much farther away. And as the Moon continues to orbit the Earth, occasionally it passes in front of the Sun.

And that, in a nutshell, is why we have solar eclipses.

ECLIPSE

VOYAGE TO DARKNESS AND LIGHT

Chapter One:
How Eclipses Work

Remember that old riddle about the tree falling in the forest? If no one is there, does it make a sound? I like to apply that to eclipses, especially the effects that the 53 eclipses I've seen have had on me. I've always felt that we, as observers, are vital parts of these events. So suppose they gave an eclipse, and no one came?

As a scientific event, an eclipse is a comic coincidence, a curiosity. Planets don't crash into each other, and stars do not explode. But put yourself into that event, and an eclipse can have a most powerful effect. Even a barely noticeable penumbral lunar eclipse does that to me. The Moon's supposed to be full, but as its brightness dims and the rays stretching away from the craters Tycho and Copernicus become so much more prominent, I become aware that inexorably, the Moon is passing through the outer reaches of the shadow of the Earth. At the other extreme, of course, is a total eclipse of the Sun, an event that stabs like a knife to the core of my emotions. Sure, if no one sees an eclipse, then the event is nothing more than a coincidence. But those who do see it leave subtly changed, and moved by its power.

Let's look at the mechanism behind that power. A total eclipse of the Moon happens here on Earth, and with planets and moons elsewhere in the solar system. As the Moon orbits the Earth once every 29½ days, it forms some angle between it, the Earth, and the Sun. Twice each month, at New or Full Moon, that angle becomes a straight line. If it precisely straight, then an eclipse takes place. Eclipses of the Sun or Moon can occur only when the Sun, Earth and Moon are exactly lined up. This can occur during Eclipse seasons, which happen twice each year. The simple geometry of planetary bodies orbiting one another in space is common enough, and because the solar system is essentially on one plane, like a record or disc, these lineups occur frequently. But on these other worlds, the effect is far less dramatic. Mars, for example, has lineups when either of its tiny moons, Phobos and Deimos, pass in front of the Sun. But these moons are so small that they produce virtually no effect at all—if you were standing on the surface of Mars, you might barely detect such an event in progress as a tiny dot crossing the Sun's surface. Jupiter's moons would, on the other hand, block out the Sun so completely that they would cause several hours of darkness. But Earth's single Moon is now at just the right distance from us that it appears to be the same size as the Sun. The result is an exquisite blocking of the Sun's bright surface, revealing the prominences of the Sun's inner atmosphere, or chromosphere, and the pearly outer atmosphere, or corona. The fact that we get to see the prominences and the corona at all is a miracle in itself, for the Moon is slowly moving farther from Earth. As it continues to recede,

The Earth's shadow envelops the Moon. Photo by Roy Bishop on January 21, 2000, from Maktomkus Observatory.

its apparent size will shrink, and it will no longer be able to completely blot out the Sun.

ECLIPSES, TRANSITS, AND OCCULTATIONS

Technically, the event called an eclipse takes place only when the body being covered is the same apparent size as the body doing the covering. A *transit* takes place when a small body passes in front of a larger one, as when Mercury or Venus pass in front of the Sun. An *occultation* is the term we use to describe the passing of a large body in front of an apparently smaller one. The Moon frequently passing in front of a star is an example.

ANNULAR ECLIPSES

As the Moon continues slowly to move away from Earth at a rate of one and a half inches each year, it will eventually seem to take up less space in the sky. Some 620 million years in the future, there will come, a day, sadly when we see the last total eclipse. The Moon will pass in front of the Sun, covering it entirely for a fraction of a second, and when that faroff event is over, total eclipses of the Sun will be a thing of the past. Even now, almost half the times that the Sun is centrally covered by the Moon, the Moon is near the farthest point of its orbit around the Earth. At that distance it does not cover the Sun completely; the result is an *annular*, or *ring* eclipse, at the middle of which the Moon is surrounded by a ring of bright sunlight.

In May 1984, I traveled to New Orleans to see an annular eclipse. A cold front had passed through the night before,

Moonrise, seven hours prior to total eclipse. January 20, 2000. Note the Earth's shadow projected onto the sky below the Moon, and ice blocks stranded by low tide. Photo by Roy Bishop, from Evangeline Beach, Nova Scotia.

leaving the usually humid city dry and clear. At the midpoint of the eclipse, the Moon's shadow swept out of the sky and almost entirely enveloped us. Overhead, what was left of the Sun shone as the thinnest of rings. A second later, the darkness whooshed away, leaving a thin crescent of sunlight. On the way home from this particular ring eclipse I stopped by to visit Clyde Tombaugh, the discoverer of the planet Pluto, and an old and close friend. "How was the eclipse?" he asked. "Did you have a ringside seat?"

Clyde got the chance to live down that pun. Ten years later a second ring eclipse took place directly over his house. Although poor weather was forecast, the day dawned bright and clear and we saw a magnificent ring. The Moon was close to apogee, and took up considerably less space in the sky than the Sun. Thus, the ring was bigger, and it lasted 11 minutes!

THE ORBIT OF THE MOON

The moon circles the Earth once every 29½ days, a period loosely coinciding with a month; the word "month," in fact, is derived from old German for "moon." The Moon's orbit is the only determinant of the Moslem calendar, and is the prime base for the Jewish calendar, whose months, though timed to the Moon's orbit, are kept in step with the solar year by the occasional addition of a leap month.

Does the Moon's phase affect human behavior? There has been much conjecture and debate on that question. Anecdotal evidence does support the idea that people tend to get rowdier around the times of full Moon. But as to whether the Moon has an effect on the Earth, the answer is absolutely clear. When the Moon is near perigee (its closet approach to Earth), its gravity causes an increase in the strength of the tides. And when Moon and Sun are approximately lined up with the Sun, tides increase.

THE NODES

Eclipses happen because of the relation between the orbit of the Earth around the Sun, and the Moon's orbit around the

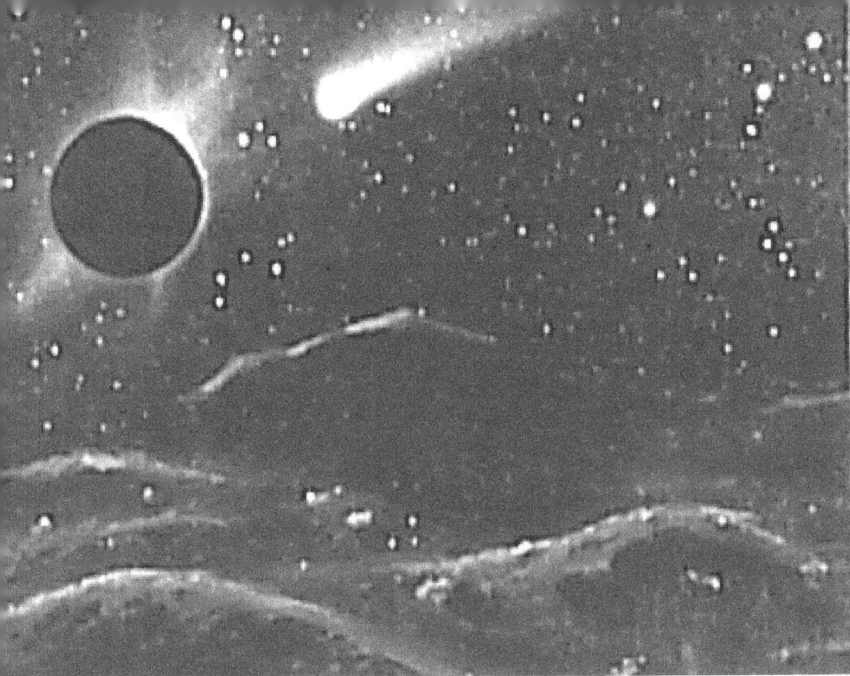

An eclipse in space, and a comet near the Sun. Original Painting by James V. Scotti.

Earth. The Earth orbits the Sun in a near-circle once every year, and the Moon orbits the Earth, reaching the Sun's position, every 29½ days. The two orbits are tilted relative to each other; if they were not, eclipses would happen every two weeks, a solar eclipse every new Moon and a lunar eclipse at every full Moon. Instead, the Moon spends part of its month-long orbit below the plane of the Earth's orbit around the Sun, and part of it above that plane. Twice each month the Moon crosses the plane of Earth's orbit. The two points of crossing, or intersection, of the two orbits are called *nodes*.

The Moon crosses a node twice each month. If the Moon is moving northward in its orbit, it's called the *ascending node*; if it is going south, it crosses the *descending node*. The node crossings take place at different phases of the Moon each month. Now, in addition to the Moon's orbit of the Earth, the Earth travels around the Sun, so the Sun appears to cross one of the nodes twice each year. When that happens, an eclipse can occur.

ON THE MOON

When eclipses occur on Earth, does anything happen on the Moon? During a lunar eclipse, the entire Moon is bathed in the shadow of the Earth, which means that a person or a camera on the Moon should witness an eclipse of the Sun (by the disk of the Earth). On April 24, 1967, humanity saw a total eclipse of the Sun by the Earth through the eyes of the U.S. space probe called Surveyor 3. The spacecraft's camera took two sets of pictures of the event. The exposures revealed that an eclipse of the Sun by the Earth, seen from the Moon, is less spectacular than one of the Sun by the Moon. From the Moon, the Earth takes up almost four times as much sky as the Moon does as seen from Earth. Thus, at the middle of the eclipse the Earth would cover the Sun and most of its corona. What part of the outermost corona was left might have been recorded by the Surveyor camera, but the exposures were not long enough to reveal any of it. At the same time that Surveyor was working to take those pictures of a solar eclipse, a lunar eclipse was taking place on Earth that Passover night.

(Since they always occur at full Moon, Passover Seders often coincide with eclipses of the Moon. I remember leaving our Seder to observe that particular eclipse.)

AN ECLIPSE OF THE EARTH

What would an eclipse of the Sun by the Earth look like from the Moon? Because the Moon's shadow is so much smaller than the Earth's, the effect for lunar inhabitants would be barely noticeable. The Earth, visible in the lunar night in full phase, would dominate the sky. If you looked closely, you could watch a tiny patch of darkness, surrounded by a fainter outer shadow, wend its way across the Earth's surface, crossing land and sea, from one side of the planet to the other, in about two hours. Truly, the cosmic pinball game of Earth, Moon, and Sun is something to see from any vantage point.

Chapter Two:
Childhood Impressions of a Darkened Sun

A few hours ago, I completed observing session No. 11619 from our own Jarnac Observatory. The sky was warm and clear this May morning as I turned my telescope to the southeast to begin my pastime of searching for comets. As I moved the telescope from one field of view to the next, and on to the field after that, I searched a pattern of sky that might reveal a new comet. There were none this time, but as I moved the telescope closer to the southeastern horizon, I thought of another observing session that took place almost exactly forty years earlier.

Now a distant memory, that observing session was No. 1 in my record book, my first formal, recorded observing session:

*1S. October 2, 1959. Partial Solar Eclipse. Just last part observed because of clouds.

OCTOBER 2, 1959

Something magical happened early that morning as a partial eclipse of the Sun worked its way over my childhood home in Montreal. Far to the south, the full darkness of the Moon's shadow cast its spell over a thin path through Massachusetts, and out over the Atlantic. But I didn't care about that. For me at age 11, all that mattered was that I was to see my first eclipse. The event was to last until 8 A.M., so that it would be conveniently over in time for the start of my day in sixth grade.

Our observing team consisted of my mother, my brother Gerry, and me. As we drove to the Mount Royal Lookout facing Montreal's east side, we worried that clouds would prevent our seeing the eclipse. We waited for a while as the Sun rose behind the overcast. Then Mother noticed that the clouds were clearing from the west, so if we moved to another site farther west, we might get to see the end of the eclipse as the Sun broke through the clouds. We sped to the new site, got out of the car, and waited along with a large crowd that had gathered there. The sky grew brighter until the Sun peeked its way through the clouds. As the Sun rose further, its crescent shape showed itself at last. It was my first eclipse, my first observing session, and I was in heaven.

JULY 20, 1963

After that early eclipse, I looked up everything I could about eclipses. I used a book that contained a map of the world crisscrossed with thin lines that curved their way across the globe. I noticed that one of those lines showed the path of an eclipse that would cross Canada and rush near Montreal on July 20, 1963. It was a curious thing for me to see that the track of not just a partial, but a *total eclipse* would be only a two-hour drive from downtown Montreal.

Reprint of Montreal Star *photograph of partial eclipse of the Sun, October 2, 1959. The picture shows the eclipse as my family viewed it that morning, and the year 1959 can be faintly seen after I wrote it in pencil that day. I also wrote that* Rosh Hashana *in 1959 was taking place, as it always does, on the day of a new Moon. Photograph by Mac*

The summer of 1963, however, found me far from my Canadian home. My address was Denver's Jewish National Home for Asthmatic Children, a spot from which I would see, at maximum eclipse, a half-covered Sun barely more exciting than my view from 1959. Since it was the policy of the asthma home not to allow their patients to return home during their year-long-plus stay, my chances of seeing this eclipse seemed remote. But Mom and Dad put in a special request that I be permitted to return home for just one week to see the eclipse. The administration of the asthma home knew that under the clearer skies of the American west, my harmless little hobby had flowered into an all-consuming passion. I used every opportunity to observe the Sun by day, projecting the solar image on a piece of paper to count and draw the sunspots. Knowing how serious I was about observing the total eclipse, the asthma home staff permitted me to travel east.

The evening before the eclipse, we had dinner with our relatives and close friends Leo and Leona Kirschberg. Observing the total eclipse was high on our minds, but Leo, an opthamologist, was adamant about taking care not to damage our eyes. He pointed out that during the partial phases of an eclipse, the Sun's visible light output drops, so that we are able to gaze at the Sun for longer periods without squinting. We also wanted to look at the changing Sun as the Moon covered more and more of its face. However, Leo warned, the ultraviolet rays coming from the Sun's photosphere are just as strong during an eclipse as they are at any other time, and these rays can permanently damage the retina.

I understood Leo's wisdom, but in reply I claimed that while this is true for the eclipse's partial phases, it is not true when the entire Sun is covered by the Moon-the total eclipse, and by far the most interesting time to see the Sun. "But how," my father asked, "can one be sure that the entire Sun is covered?"

"The darkness is supposed to whoosh in like . . . like . . ."

"Like a rapidly approaching thunderstorm?" Leo asked, helpfully.

"Much faster!" I said, having no understanding whatever

July 20, 1963. My first total eclipse was a lucky one—A minute after totality was over, thick clouds covered the Sun for the rest of that memorable day near Plessisville, in southwestern Quebec. Constantine Papacosmas took this photograph a short distance from where I watched.

of how vastly different the next day's event would be. In any case, we weren't sure we'd see this eclipse at all. "Chances for viewing tomorow's eclipse in southwestern Quebec," the weather forecaster intoned, "are poor." Eclipse day dawned mostly cloudy, but we did see the Sun break through a few times while driving to our carefully selected site at Plessisville. Our group of four—Mom and Dad, my friend Paul Astrof, and I—were on our way to be a part of a four-way lineup that included the Sun, Moon, Earth, and us.

The thing I remember most from the following day was Dad's astonished reaction to the fact that the eclipse started right on time. To astronomers used to traversing the globe to see an eclipse, that innocuous first contact of Moon and Sun is taken for granted. Dad was amazed that a centuries-old prediction of some long-gone astronomer was coming true before his eyes.

He was not a scientist, but Dad was good at seeing the poetry around him. As we prepared to watch the eclipse, we knew that at that moment thousands of other teams would be watching the eclipse as it tracked across the continent, from Alaska, all the way to us, in under two hours. Just at that moment another interesting team was watching the last thin crescent of the Sun disappear as the Moon, over Alaska, pushed it into total eclipse. Braving mosquitos and a dismal weather forecast, that team consisted of planetary astronomers Brad Smith, Clyde Tombaugh, and Clyde's wife Patsy. The weather cleared in time for them to view a beautiful total eclipse. Clyde was already famous as the discoverer of Pluto, the ninth major planet of our solar system. Brad would become well known two decades later as the leader of Voyager's imaging team, a project that launched two intrepid spacecraft to explore, for the first time, the outer worlds of our solar system.

Clyde and Brad wanted everything observed, and timed, as accurately as possible. Accordingly, Clyde asked Patsy to keep her eyes on the stopwatch and count the 90 seconds of total eclipse. Obediently, Patsy did, and thus only got two quick glances of this eclipse. It was a memory she would carry until she finally would see her next total eclipse 35 years later.

The sky was clear in Alaska that eclipse morning, but late in the afternoon, in southwestern Quebec, it was mostly cloudy. The Sun broke through occasionally, but as it went deeper and deeper into partial eclipse, heavy clouds on the western horizon seemed sure to block our view of the totality. In the last few darkening minutes before the onset of total eclipse, Dad looked up. "Come on, please give my son a break," he said quietly, "a break in the clouds."

Somehow his prayer was answered. The clouds held back to reveal the incredible sight of the onrushing shadow and a faint circular corona. With the sunspot cycle nearing its 11-year minimum, we did not expect to see any prominences. A minute later, the shadow lifted and a thin crescent of sunlight reestablished itself. Stunned and excited by what we saw, we just stood there. And then the clouds came, covering up the rest of the eclipse. But it didn't matter; we saw what we had come to see.

As we prepared to drive back to Montreal, the Moon's

shadow continued its trek across southern Quebec and Maine, and out into the Atlantic Ocean. Near the end of its journey, it covered a swath of ocean some 210 miles (350 kilometers) southeast of Halifax. As we drove back that cloudy and happy afternoon, I had no idea that, 36 summers later, the Moon's shadow would again be hitting the ocean at the same spot, that I would be there, and that I would see that eclipse again.

Chapter Three:
Of Cycles and Friends

I saw the 1963 eclipse on July 20th—for the first time. Although I didn't know it at then, if those clouds came in too soon I would have the chance to see the same eclipse again, on July 31, 1981, if I cared to travel to the Soviet Union. Eighteen years and a few days later, the Moon's shadow touched down a third of the world away, and crossed over the Soviet Union. I didn't see the eclipse then, but my colleague and friend Bart Bok was able to, thanks to an arrangement I had made for him through a mutual friend. A famous specialist on our Milky Way galaxy, Bart used the trip to see the eclipse *and* review his favorite topic—the "bigger and better Milky Way"—with his Soviet colleagues. A fierce promoter of international cooperation in science, he also used the trip to try to strengthen the bonds of cooperation during that dangerous Cold War period.

I missed the eclipse that time. But another eighteen summers later, the shadow would sweep out of the sky once more, touching the Earth in the Atlantic Ocean some 210 miles southeast of Halifax, Nova Scotia. The date: August 11, 1999.

SAROS 145: AN ECLIPSE THROUGH TIME

The same eclipse, over and over again. Each eclipse repeats itself every 18 years, $10\frac{1}{3}$ days (or $11\frac{1}{3}$ days, when we take leap years into account.) The eclipses are almost identical, except that thanks to that extra $\frac{1}{3}$ day, they fall over a different part of the Earth. We call each of these cycles of repetition a *saros*, which derives from a statement from an ancient astronomner named Suidas, that the length of the saros was $18\frac{1}{2}$ years. *Saros* is the classical Babylon term for the number 3600, and hence refers to their period of 3600 years. However, to the ancient Chaldeans and Greeks, the saros had a different meaning—the period of 18 years, 10.3 days (or $6585\frac{1}{3}$ days) during which eclipses would repeat themselves.[1] The Chaldeans discovered that $6,585\frac{1}{3}$ days after new Moon at one of the nodes, the Moon has orbited the Earth precisely 223 times, and returned to the new Moon phase at the same node. At the same time, the Sun, in its apparent journey around the ecliptic, has passed through 19 eclipse years (of 346.62 days), it has also returned to the same node, and another eclipse occurs.

1. *OED, Oxford University Press, 1971.*

Some terms:

Ecliptic: The Sun's apparent path through the sky each year, as seen from Earth.

Moon's ascending Node: The place in the Moon's orbit around the Earth where it crosses, moving north, the orbit of the Earth as it travels round the Sun.

Moon's descending Node: The place in the Moon's orbit where it crosses, moving south, the orbit of the Earth.

Eclipse season: A short interval of time, lasting about a month, in which the Sun's apparent motion along the ecliptic places it close to one of the Moon's nodes.

Eclipse year: The time for the Sun's apparent motion to carry it from one ascending node of the Moon back to the ascending node. Eclipse years last 346.6 days.

Saros 145, for example, is the series of eclipses that includes the event of August 11, 1999. It began on January 4, 1639, with a tiny partial eclipse near the North Pole. Every 18 years, 10⅓ days, the partial eclipse repeated, a bit farther south, and with a bit more of the Sun covered. On June 6, 1891, Saros 145 produced its first eclipse in which the moon passed centrally across the sun, an annular eclipse in Siberia.

The Moon was not large enough to cover the Sun, so the Sun appeared as a ring of fire about the darkened Moon. Eighteen years later, on June 17, 1909, an eclipse that was annular at each end of the track but briefly total in the middle, raced over Greenland and central Russia.[2]

BART BOK AND THE SAROS'S FIRST TOTAL ECLIPSE

On June 29, 1927, Saros 145 produced its first completely total eclipse on a track that stretched through Scandinavia and over the Arctic. That was the eclipse that a young Bart Bok watched with his friend Gerard Kuiper. As we have seen, Bok would become famous with his work on the Milky Way, and Kuiper would be famous with his research on the eponymous Kuiper Belt of comets beyond Neptune. As young scientists freshly graduated from the University of Leiden, the two friends set out on a bicycle ride with tents and cookery. Their bicycles carried them on a journey from Holland, through Germany, and to Hamlet's castle at Elsinore in Denmark. From Oslo they pedaled to the northwest over mountain passes and, finally, near the end of June, they arrived at the Hallingdal River in time for the eclipse. They had designed an experiment to determine the true color of the Sun's corona by shining lights on disks of different colors and comparing the results with their visual observations of the corona. The test had to have been set up so that the lights pointed away from the

2. Fred Espenak, *Total Solar Eclipse of 1999 August 11* (NASA Reference Publication 1398, 1997) 12. See also Mark Littmann, Ken Willcox, Fred Espenak, *Totality: Eclipses of the Sun* (Oxford: Oxford University Press, 1999), 182.

The total solar eclipse of July 30, 1981, seen from an airplane. Photo by Stephen J. Edberg, Moonshadow Expiditions. This is the repeat performance of the eclipse of 1963.

observers in order not to disrupt their observing of the corona.

Sadly, thick clouds prevented the group from seeing the 50 seconds of total eclipse and trying their experiment, but the trio did observed the rapidly darkening sky as the valley plunged into the Moon's shadow. "The shadow of the Moon came very quickly over," Bart recalled, "and we saw the darkness quickly approaching across the snow on the mountains. That was the most glorious sight."[3]

3. David H. Levy, The Man Who Sold The Milky Way: A Biography of Bart Bok (Tucson: University of Arizona Press, 1993, 11–13.)

THE SAROS GETS BETTER

When the eclipse happened again on July 9, 1945, the Moon's shadow hit the Earth in the Western United States and climbed through central Canada, Hudson's Bay, Greenland, and northern Europe. Eighteen years later, Saros 145 performed again, this time on July 20, 1963, for Clyde Tombaugh and his friends, and for me and my family, on opposite ends of the North American continent. On July 31, 1981, an older Bart Bok saw the eclipse for the second time in Russia. On August 11, 1999, Patsy Tombaugh and my family were reunited to view the eclipse.

On August 21, 2017, the Saros will perform again, this time for the continental United States. On September 22, 2035, there will be a total eclipse over Asia and the Pacific, and another, over Northern Africa and the Indian Ocean, on September 12, 2053. The series will reach its middle point with a four-minute eclipse on February 25, 2324, but the length of totality will increase until the eclipse of June 25, 2522, when the Moon will cover the Sun for 7 minutes, 12 seconds. The final total eclipse, far in the southern hemisphere, occurs on September 9, 2648. The last of Saros 145's 77 eclipses will be a small partial eclipse at the south pole on April 17, 3009.

SAROS 145 AS AN EXELIGMOS

Because of that third of a day that helps define the saros, each successive eclipse in a saros cycle takes place a third of the way around the Earth. It is interesting that the tail end of the 1963 eclipse crossed over the same stretch of water on which we saw the beginning of the 1999 eclipses two cycles later. The 1999 eclipse was two-thirds of the way across the surface of the Earth from 1963, so it makes sense that its beginning would cross over the end of the path of the earlier eclipse. The path of the 2017 eclipse, however, returns to the same part of the Earth as 1963, just further south. This triple-saros cycle is also called an *exeligmos*. Bart Bok was lucky enough to witness an *exeligmos* by seeing the eclipse of June 29, 1927, from Scandinavia, and the one of July 31, 1981, as well as the one in 1999.

THE METONIC CYCLE

Although the 1963 eclipse and the 1999 eclipse were the same one according to the saros cycle, they were not according to the Metonic cycle. Since 430 B.C., when the Greek astronomer Meton discovered it, we have known that the sequence of phases of the Moon are the same every 19 years. There was a new Moon, and an eclipse, on February 26, 1979, one I saw in full glory near Winnipeg, Manitoba. When I began planning to see the February 26, 1998 eclipse in the Caribbean, I recalled the earlier one. The new Moon on October 2, 1959—my first eclipse—was followed by another new Moon and a partial eclipse in the eastern hemisphere, on October 2, 1978. However, this cycle is not quite as regular as the Saros, since it skips one of every five repetitions. So there was not an eclipse on October 2,1997. Also, unlike the saros, the metonic cycle can't be used to predict where an eclipse will take place.

A PERSONAL CYCLE

For me, eclipse cycles are not just science coincides; they are very personal. The 1999 eclipse was a strong reminder of the roles that Bart Bok and Clyde Tombaugh, had on me. I met them both at a meeting in Tucson during the summer of 1980. That meeting led to two biographies, one that I wrote on Clyde in 1991,[4] and one about Bart in 1993. But the most profound memory still comes from the 1963 eclipse, which I shared with my father, and comes from my Dad, who was so excited when the 1963 eclipse started right on time—as though some great celestial clock had been set correctly. My thoughts often go back to that eclipse of long ago, and how I wish that Dad were here to see another one with me.

4. Levy, *Clyde Tombaugh: Discoverer of Planet Pluto* (Tucson: University of Arizona Press, 1991).

As I planned to travel to the Caribbean to view the February 26, 1998 eclipse, I was reminded of the eclipse 19 years earlier, on February 26, 1979. Photo by David H. Levy.

Chapter Four:
Sun, Moon, and Surprise

W hen the Moon passes in front of the Sun, cutting off its light, the result is one of the most spectacular sights Nature has to offer. Day turns into twilight as the shadow of the Moon cuts across land and sea, and the Sun appears as a jeweled crown hanging in the sky. But if all this isn't enough, the alignment of Sun, Moon, and Earth has other effects as well.

Twice each month, whether or not an eclipse takes place, the Sun, Earth, and Moon are aligned in such a way that tides on Earth are stronger than usual. Although this effect is most noticeable with the ocean tides, it is felt by the Earth's rocky crust as well. It might be just a coincidence that the great Tokyo earthquake of September 1, 1923, in which 140,000 people were killed, took place just 5 days after the August 26 partial eclipse of the Moon, and the Turkey earthquake of 1999 occurred just five days after the total solar eclipse that is the subject of this book passed over the same location.

ASTRONOMICAL TIDES IN THE MINAS BASIN

The higher than normal tides that accompany these align-
ments of Sun and Moon are called astronomical tides, and if
a low pressure area is centered over the area at the same
time, they can be truly remarkable. On the shores of Nova
Scotia's Minas Basin, these tides are amplified by the unusual
resonance effect of the Bay of Fundy. Resonance works like
as a child on a swing, which can be hurled higher and higher
if someone pushes the swing each time it reaches a limit.
These tidal effects are enhanced when the Sun and Moon are
pulling together. On January 29, 1979, physicist Roy Bishop
of Acadia University compared low and high tides at
Hantsport, Nova Scotia, on a day near new Moon (in fact a
month before the February 26 solar eclipse), and on a day
when a low pressure system was passing through the area.
His goal was to obtain photographs from the same location,
the edge of the Hantsport pier, at low and high tide. The low
tide photograph was easy enough to take; he drove his car
out to the edge of the pier (he wanted the car in the picture
for scale), then he walked back to take the picture. The
wooden mounts for the pier were fully visible in the mud
that cloudy morning. When he returned six hours later, he
was astonished to find—nothing. The mounts for the pier
were gone, buried under water, as well as the pier itself! The
water level had risen by more than 50 feet. Roy couldn't
drive his car onto the pier; in fact, to work his way to the
same spot he had to creep, step by careful step, across the
submerged planks, snap the second picture, then carefully

work his way back. The result is a graphic demonstration of the influence of the Moon's gravity.

SURPRISING SCIENCE DISCOVERIES

When the Moon eclipses the Sun, Earthbound astronomers have a golden opportunity to study the star that gives us life. Since eclipses provide the only opportunities to see the Sun's corona from the surface of the Earth, scientists study the corona during eclipses. They try to understand why it is so hot; far hotter than the surface of the Sun, the corona shimmers at some 2 million degrees Celsius. We study the corona to learn how it affects the Earth as well—in fact, its outer reaches are the solar wind of radiation that flows past the Earth.

The Eclipse of August 18, 1868, excited several groups of scientists. On that day a group including John Herschel, great-grandson of William Herschel, discovered that solar prominences are composed of hydrogen gas. At that same eclipse, J. Norman Lockyer and Pierre Janssen detected a yellow line in the spectrum of the Sun's corona that signified a new element, which he named Helium, a quarter-century before it would be detected on Earth.

Comets have even been found during total solar eclipses. Arthur Schuster photographed a new comet during the total eclipse of May 17, 1882.[1] People enjoying the eclipse of November 1, 1948, to cite a recent example, were stunned to

1. Mark Littmann, Ken Willcox, and Fred Espenak, *Totality: Eclipses of the Sun* (New York, Oxford: Oxford University Press, 1999), 157.

see a comet brighter than Jupiter about a Moon diameter from the eclipsed Sun, and with a tail that stretched toward the horizon![2]

The eclipse of June 8, 1918, became famous for something else that happened on the same day. As comet hunter Leslie Peltier wrote in *Starlight Nights*, his biography:

> Along the narrow track of totality astronomers from all over the world packed up their precious plates and prepared to leave for home. Weeks before, they had assembled here and had carefully taken their places in line in order to see a spectacle that would last just two brief minutes. For the most part they left well pleased with the performance though, as always, some had been unfortunate in their choice of seats along the lengthy aisle. And as they started homeward not one in all that far-flung audience could know that this was just an intermission and that the show they had come so far to see would be a double feature.
>
> When darkness came that evening I clamped my spyglass to the grindstone mount, which still was standing at the station underneath the walnut tree. I hoisted it up on my shoulder and carried it out into the middle of the front yard and stood it where I would have a clear view of the variable stars in the

2. Gary Kronk, *Comets: A Descriptive Catalog* (Hillside, N.J.: Enslow, 1984), 147.

southeastern sky. That was the night that I forgot all about telescopes and variables for as I turned and looked up at the sky, right there in front me—squarely in the center of the Milky Way, was a bright and blazing star![3]

The discovery of an exploding sun, or nova, is still one of the most exciting things that can happen in astronomy, and it was an incredible thing that such a star's light would have traveled for thousands of years through space to arrive on Earth's doorstep just as the Sun was being eclipsed.

3. Leslie C. Peltier, *Starlight Nights: The Adventures of a Star Gazer* (1965: Cambridge, Mass.: Sky Publishing Corporation, 1999) 94.

Chapter Five:
The Power of Gravity

T he Moon's gravity, aided somewhat by the gravitational pull of the Sun, is strong enough to affect the tides on Earth. The gravity of the much larger Sun is strong enough to control the orbits of the planets, including Earth and Moon, and the eclipses they produce. And according to Einstein's theory of relativity, a beam of light from a distant star will be bent by the curvature of space as it passes near the Sun.

What does this bending of light have to do with eclipses? The answer is a story that began in the 1890s, when a teenaged Albert Einstein had a thought. "What would the world look like," he mused, "if I rode on a beam of light?" The answer, he later figured, was that the planet would be frozen in time, its clocks still, its action caught as in a photograph. A decade after he first thought about that, the young physicist, then unable to find a job in physics, began work at the Swiss patent office. Einstein's responsibilities there were not particularly time consuming, and he had time to ponder questions about physics, like the relationship between matter and energy.

The result was Einstein's special theory of relativity,

which appeared in a 1905 article without any reference or citation—virtually unheard of for a research paper. This was a completely original piece of work. Later that year, Einstein attached an additional thought to that article, the tiny equation $E=mc^2$. In those simple letters lay the idea that mass and energy are equivalent. Late in 1915, Einstein's general theory of relativity offered a new definition of gravitation that related it to space and time. In Einstein's physics, gravity is not a force but rather geometry. As any object moves, whether it is a baseball, a planet, or a star, it follows a geometric path shaped by the unified effect of mass and energy. Newton invoked a force of gravity to make his Universe work, and in almost all cases, Newton's laws fit. But where there is a lot of matter, like a star, Newton's laws fail, and it is important to see gravity not as a force of Newton but as Einstein's geometry of space and time.

How could this beautiful, simple relation be tested? In the years between 1905 and 1919, Einstein's theory caught the attention of physicists, but many thought that it was unprovable. Such a theory would be forever relegated to the backwaters of physics as a curiosity. In 1918, the British astronomer Arthur Eddington wrote the first English-language account of relativity, and he noted how the theory was successful in solving the old problem about Mercury's orbit. As Mercury circles the Sun at a distance of only 36 million miles, its perihelion, or closest point in its orbit to the Sun, shifts a small amount with each orbit. The entire orbit is precessing by 43 seconds of arc per century, a tiny amount to be sure, but one which Newton's theory of gravitation could not explain. If

Newton's laws were right, another planet closer to the Sun must be affecting Mercury's orbit. Scientists searched for such a world—and even named it Vulcan—for decades. There is no such planet. Instead, Newton's model of gravitation turned out to be insufficient. Einstein's theory explained this shift.

THE ECLIPSE OF MAY 29, 1919

Eddington noted that Einstein's theory "further leads to interesting conclusions with regard to the deflection of light by a gravitational field," and that it could be tested through an experiment.[1] It seemed that in 1918, Eddington was sure just what experiment would work. By photographing a star near the Sun, and then comparing its position with that on other photographs taken when the star is far from the Sun, Einstein's relation could be tested. Although this proposed test is fine in theory, when the star's light passes that close to the Sun, we shouldn't be able to sight the star at all because of the sun's overwhelming luminosity. Except, that is, during an eclipse.

Just what was this result? Eddington needed to answer two questions. First, does light have weight, as Newton suggests? And if the answer is yes, Is the amount of deflection of the star's perceived position in agreement with Newton, or with Einstein? The deflection, incidentally, is in the opposite direction from the body doing the deflecting. The star will appear to be displaced outward, or away from the Sun, by the same

1. A. Vibert Douglas, *The Life of Arthur Stanley Eddington* (London: Thomas Nelson and Sons, 1956), 39.

amount as the total deflection.

For the experiment to work, Eddington needed to use distant stars, not planets, asteroids, or comets within the solar system. The bending of the star's light is detectable only with stars apparently near the Sun, and these stars can be seen only during a total eclipse. But the Sun's corona is also bright, so the only eclipse that will work is one when the Sun is near a group of moderately bright stars. It turned out that just such a coincidence was about to happen, and Eddington developed a plan to photograph stars in the Hyades star cluster, from Principe, a small island off Africa's west coast, which, on May 29, 1919, would be under the darkness of a total eclipse of the Sun. The cluster would be just south of the eclipsed Sun, its stars bright enough to be captured on the photographic films of the time.

The displacement of a star close to the Sun is measured in comparison with stars that are farther from the Sun and not displaced. In order to measure displacement, astrometrists (astronomers who measure the positions of stars) therefore need to observe stars close to and farther from the Sun. "In a superstitious age," Eddington wrote, "a natural philosopher wishing to perform an important experiment would consult an astrologer to ascertain an auspicious moment for the trial. With better reason, an astronomer to-day consulting the stars would announce that the most favorable day of the year for weighing light is May 29. The reason is that the sun in its annual journey round the ecliptic goes through fields of stars of varying richness, but on May 29 it is in the midst of a quite exceptional patch of bright stars—part of the Hyades—by far the best star-field encountered. Now if this problem had been put forward at some other period of history, it might have been necessary to wait some thousands of years for a total eclipse of the sun to happen on the lucky date. But by strange good fortune an eclipse did happen on May 29, 1919. Owing to the curious sequence of eclipses [the Metonic cycle we have already discussed] a similar opportunity will recur in 1938; we are in the midst of the most favorable cycle. It is not suggested that it is impossible to make the test at other eclipses; but the work will necessarily be more difficult."[2]

Planning an expedition to Africa to observe an eclipse in order to confirm the ideas of a German scientist was an almost insurmountable problem, but Sir Frank Dyson, then England's Astronomer Royal, was able to persuade the government that this eclipse presented a rare opportunity indeed, and that they should spend 1000 pounds for an expedition to test the theory of relativity.

AN AMAZING EXPEDITION

Five months before the expedition began, Eddington photographed the Hyades field, using the same telescope as would be brought to Africa. With the Hyades far from the Sun, this photograph would serve as a base for comparison. As summarized in A. Vibert Douglas's biography of Arthur Stanley Eddington, the night before the expedition set sail there was a discussion about just how much deflection the

2. Sir Arthur Eddington, *Space, Time, and Gravitation: An Outline of the General Relativity Theory* (Cambridge: Cambridge University Press, 1920), 113.

star would suffer. If the deflection was a tiny 0.87 arc seconds, then it would confirm Newton's classical theory of gravitation. If it were much greater than that, or 1.75 arc seconds, then Einstein's theory would be confirmed. On the evening before the sailing began, Cottingham, who to accompany Eddington, asked jokingly what would happen if the star's deflection was double what Einstein had predicted. Dyson, Britain's royal Astronomer and the man who had planned and arranged funding for the expedition, replied, "Then Eddington will go mad and you will have to come home alone!"

Eddington described this fateful expedition in his notebook:

"We sailed early in March to Lisbon. At Frunchal we saw [Davidson and Crommelin, the other expedition] off to Brazil on March 16, but we had to remain until April 9 . . . and got our first sight of Principe in the morning of April 23 . . . about May 16 we had no difficulty in getting the check photographs on three different nights. I had a great deal of work measuring these."

The group arrived in Principe with a 13-inch diameter, 11-foot, 4-inch long refractor. They stopped the lens down to 8 inches to improve the sharpness of its images. The telescope was mounted in a fixed position, and a mirror, or coelostat, directed the light from stars into the telescope.

"On May 29 a tremendous rainstorm came on. The rain stopped about noon and about 1:30 when the partial phase was well advanced, we began to get a glimpse of the sun. We had to carry out our programme of photographs in faith. I did not see the eclipse, being too busy changing plates, except for one glance to make sure it had begun and another half-way through to see how much cloud there was. We took 16 photographs. They are all good of the sun, showing a very remarkable prominence; but the cloud has interfered with the star images. The last six photographs show a few images which I hope will give us what we need. . . ."[3]

A year later Eddington expanded on the details of those few minutes of total eclipse:

"There was nothing for it but to carry out the arranged program and hope for the best. One observer was kept occupied changing the plates in rapid succession, whilst the other [presumably Eddington himself] gave the exposures of the required length with a screen held in front of the object glass to avoid shaking the telescope in any way.

For in and out, above, about, below,
'Tis nothing but a Magic Shadow-show
Played in a box whose candle is the Sun
Round which we Phantom Figures come and go.

Our shadow-box takes up all our attention. There is a marvellous spectacle above, and, as the photographs afterwards revealed, a wonderful prominence-flame is poised a hundred thousand miles above the surface

3. Douglas, 39.

of the sun. We have no time to snatch a glance at it. We are conscious only of the weird half-light of the landscape and the hush of nature, broken by the calls of the observers, and beat of the metronome ticking out the 302 seconds of totality."[4]

Eddington's 16 exposures ranged in time from 2 to 20 seconds. The first ones did not record any stars, but they did capture the prominence. But happily the clouds cleared more toward the end of totality, and one photographic plate recorded five stars. Once all the pictures had been processed, Eddington made his first measurements at the eclipse site a few days after the event. The two photos were placed "film to film" in the measuring machine so that the star images were close to identical. "In comparing two plates," Eddington wrote, "various allowances had to be made for refraction, abberation [of the telescope lens], plate-orientation, etc."[5]

Again from Eddington's log: "June 3. We developed the photographs, 2 each night for 6 nights after the eclipse, and I spent the whole day measuring. The cloudy weather upset my plans and I had to treat the measures in a different way from what I intended, consequently I have not been able to make any preliminary announcement of the result. But the one plate that I measured gave a result agreeing with Einstein."[6]

As he completed the reduction of this plate, Eddington realized the significance of his result. Turning to his col-

4. Ibid., 114–115.
5. Eddington, 114–115.

league, he smiled and said, "Cottingham, you won't have to go home alone." They packed their precious plates and returned to England. Four additional plates, of a different type that could not be developed in the hot African climate, were developed there, and one of them confirmed the result shown on the first successful plate.

For a test as crucial as this one, Eddington had to make sure that instrument errors could not have led to the result. As a check, Eddington photograped a different star field, at night, with his arrangement at Principe; the field was photographed from England. If the "Einstein deflection" were the result of a telescope error of some kind, it would have turned up in these check plates. But no changes were found in the stars on the check plates.

The Brazil part of the expedition had much better luck. Their weather on eclipse day was surperb, and they remained at their site for two additional months in order to photograph the same region of sky under cover of morning darkness. They had two telescopes, one similar to the African scope, and a much longer, 19 foot long, 4-inch diameter refractor. But despite the best preparations, few observing expeditions are free of surprise, uncertainty, and disappointment. When the Brazil expedition returned home finally, its measured deflections with the larger telescope did *not* agree with Einstein, but with Newton! This was a shocking result, one which would inevitably delay any announcement, one way or another, about relativity. Eddington suspected that the Sun's rays in the

6. Douglas, 39.

clear Brazil sky might have distorted the mirror; in this one case, bad weather helped. "... at Principe," he wrote, "there could be no evil effects from the sun's rays on the mirror, for the sun had withdrawn all to shyly behind the veil of cloud."[7]

The final verdict on relativity thus had to wait until a measuring engine could be modified to accept the seven oddly sized plates taken through Sobral's 4-inch refractor. These plates seemed ideal, and their images were perfect. So was the result they revealed: a deflection in strong agreement with the results in Africa, and one in favor of Einstein's theory.

Einstein was thrilled with this result. "I should like to congratulate you on the success of this difficult expedition," he wrote in a letter that began "Lieber Herr Eddington!" "I am amazed at the interest which my English colleagues have taken in the theory in spite of its difficulty.... If it were proved that this effect does not exist in nature, then the whole theory would have to be abandoned!"[8]

Einstein was right, and his substantiated theory of relativity made the front pages of newspapers around the world. The man in the patent office had completely redrawn our understanding of the structure of the Universe, and his idea was proved correct thanks to a total eclipse of the Sun. Perhaps it is the special magic of solar eclipses that makes us want to see something superb come out of them, something that utterly blows us away. In their discovery that the Sun's mass bent the light from nearby stars, Eddington and his col-

7. Eddington, 117.
8. Douglas, 40–41.

leagues helped Einstein give us a new universe, or at least a completely new understanding about the old one. In 1922, C.A. Chant, one of Canada's most highly regarded astronomers, mounted a most unusual eclipse expedition that lasted many weeks. With his wife and daughter he travelled by train from Toronto across Canada to Vancouver, and then by boat to Western Australia. His long journey was a success, and he confirmed Einstein's theory again with fresh measurements.

That is quite a thought to ponder when you next have the opportunity to watch an eclipse of the Sun.

Chapter Six:
The Eclipse Experience

N o matter how many eclipses I might see, I am always amazed at the uniqueness of each experience. As I was soon to learn, Bart Bok was right about seeing his 1927 eclipse-even a clouded-out eclipse is something to remember. Eclipses seen on snow, over the ocean, alone, or with others, all work their special magic on those who are fortunate enough to observe them.

1970

In the spring of 1970 I was a student at Acadia University in Wolfville, Nova Scotia. Knowing that the Moon's shadow would trace a path up the Atlantic seaboard that March 7, and pass over Nova Scotia about an hour's drive south of Acadia, I joined a group of friends to catch the eclipse's total phase. We were clouded out, but we did see the Moon's dark shadow race across the clouds.

On that same day, Roy Bishop was smarter than I was. A physics professor at Acadia, Roy was starting to expand his interest in astronomy. Within a decade he would be National President of the Royal Astronomical Society of Canada, and editor of its *Observer's Handbook*. Back in 1970, Roy also wanted to view the eclipse. On seeing the same clouds I did, he checked a weather map in the local newspaper that day and thought that clearing should be coming from the west, but a little slower than had been forecast. He turned the pages of his phone book until he found listings for a town about 100 miles to the west. He dialed the number of a flower shop, and when they answered, he said, "Hello, this is Roy Bishop of the Physics Department at Acadia. Is the Sun shining there?"

"Why, yes, it is!" came the answer. Roy's group drove out there. "Words cannot describe the beauty of the corona and darkened sky I saw," he wrote to the Canadian Broadcasting Corporation the next day. Although I missed seeing that eclipse directly, I will never forget how the landscape was thrust into a darkness more profound that at any other eclipse I have seen since.

1979

It had been many, many years since my first eclipse in 1963, and I really wanted to see another one under a clear sky. That opportunity came on February 26, 1979—the last total eclipse visible over the continental United States until 2017. I was in Canada, though, set up near near Lundar, Manitoba. The sky was supposed to be cloud-filled that day, but the night before the eclipse a weak high pressure system formed over southern Manitoba, giving us a beautiful view of the corona and several spectacular prominences.

Twenty-four years after I watched the Moon's shadow blacken the clouds during the March 7, 1970 total eclipse, Roy Bishop captured this annular solar eclipse from the same site, Crystal Crescent Beach, Nova Scotia. Photo by Roy Bishop.

1991.

In 1991, I saw an eclipse that was part of mighty Saros 136, the producer of one of the longest and finest series of eclipses ever seen. This saros was responsible for the famous Einstein eclipse of 1919, and was returning for an even better one. It was going to be a rare eclipse that crossed the site of the world's largest observatory, whose telescopes were poised to study the eclipse as never before.

In Hawaii's early morning, the total eclipse would last almost four minutes. But where I was, with a *Sky and Telescope* group in Mexico's province of Baja California, the eclipse would culminate near high noon where totality would be longest-very close to *seven minutes* of darkness!

For the first half hour it was hard to imagine that anything ununusal was taking place. The sky was still bright, and the beach was as crowded as any beach would be on a holiday afternoon. The sky did not darken linearly and steadily, as it does after sunset. My first thought that the light was changing came as I noticed that the sky was not quite as bright as it should be at high noon in summer. Through my welder's glass I looked and saw that fully half the Sun was gone. By the time the Sun was three-quarters covered, the pace of darkening was increasing rapidly. I looked away from the sea toward our hotel, which now towered into a navy-blue sky. We looked at a hundred crescents projected by spaces between leaves, by breaks in a straw hat, between the crossed fingers of two hands-the crescents were everywhere.

The sky was now darkening fast. Cameras were being reloaded in these final minutes. A man from the bar

The partial phase of the eclipse of the sun from February 26, 1979 by
David H. Levy

approached, armed with bottles. "Corona beer?" he inquired hopefully, and sold a bottle. Now there was just a sliver of Sun left, and the Moon was moving so fast I could see the Sun shrink as the seconds ticked away. Directly to the west the dark shadow of the Moon gained substance. Now the darkness was coming in waves! Was this my imagination, or was I seeing some sort of shadow band effect all around me, as the sky got two steps darker, then one step brighter, then three steps darker again. I had seen this effect in Manitoba at the 1979 eclipse.

Then I turned around and looked toward the vanished Sun.

In its place was a jewelled crown. Stretching some three solar diameters east and west, through the telescope it was rich with streamers and intricate brushes of light. On the north and south sides were a series of smaller eruptions of rays, shining outward like the mouths of baby birds in a nest.

Totality lasted long enough to allow me to use my telescope in a brief search for comets. My telescope passed over the third magnitude star Delta Geminorum, almost lost in the Sun's corona. I realized that we were witness to a strange coincidence of history: the eclipse was taking place directly over the spot where Pluto was when Clyde Tombaugh found it over sixty years ago!

Eclipses tend to heighten the senses. Listening to the sounds of the eclipse—the birds flying and strutting about, the incessant clicks of a million cameras—all accompanied the eerie light that surrounded us. Despite the fact that we were deep in the Moon's shadow, the sky was bight as twilight and landscape features were plain. Around the horizon was a bright red glow.

Total Eclipse of the Sun, February 26, 1979. Photo by David H. Levy.

Other observers reported strange environmental effects. At her site, astronomer Jean Mueller reported several cows "coming home," single file into their enclosure. (And down the beach from us, some one—of course not from our group—decided to show it all at mid-eclipse.) With the intricate structure of the corona just hanging at the zenith, time seemed to stop. However, there were complex changes occurring all around. As we grew deeper into shadow the horizons continued to darken slightly. As the Moon moved off the west side of the inner corona a glorious orange prominence appeared. Through the telescope the prominence was so brilliant it was difficult to look at, and it seemed to have erupted off the disk of the Sun, just hanging there in space.

In the final seconds of totality the western sky started to brighten rapidly. I looked back at the Sun, admiring the prominence as the western limb brightened. Suddenly a sharp speck of photosphere stabbed through the darkness, slowly spread out into a thin crescent, and the eclipse was over.

1998

Seven years later, I saw my next total eclipse, this one aboard the *Dawn Princess* in the Carribean. On February 26, 1998, this eclipse was the next one in the Metonic cycle following the one I saw in the frozen Canadian prairie in 1979. This time Wendee, my new wife, joined me in a much warmer climate.

Although I had now been under the Moon's shadow eight times, five for total eclipses and three for annular eclipses, I was still not prepared for the splendor beyond words that the experience offers. And this eclipse, far from being just

The same eclipse of February 1979, a few moments later than the frame reproduced on page 53. Photo by David H. Levy.

another eclipse, was the most precious of them all, for it was my wife Wendee's first experience of totality. The onset of darkness came gradually at first; with half the Sun gone, we realized that sunglasses were no longer necessary. But as the minutes rushed by, the gathering darkness began to hit us head-on. And as the crescent Sun shrank to a sliver, I tried to let Wendee experience as much of the event as she could.

"Glass on," I said, "now look at the Sun!" With her welders' glass, Wendee saw that the sliver was getting smaller. "Now look away from the Sun, and remove the glass." Venus had just popped out of the weird twilight that was engulfing the ship. Looking westward, I asked her to notice the distant ribbon of darkness that was the onrushing shadow of the Moon. "Glass on, and look at the Sun!" The sliver was little more than a line of light. "Glass off, look at the shadow!" In a matter of seconds, that distant ribbon was gaining strength and power. "Glass on!" The Sun was a short line. "Glass off!" The shadow was rushing at us fast. "Glass on!" The Sun was a point of light. "Now Wendee, keep looking at the Sun, and glass off." The diamond ring was beyond expression, a bright spark of sunlight and a thin corona. Within two more seconds, the diamond itself gave way to the Moon, and the corona swelled in size. Mercury, Venus, Mars, and Jupiter, all these planets hung in the sky, servants to a Sun whose light had been quenched. Our friend, Tim Hunter, watching with us, called out these planets as they appeared. For just a few minutes, time stopped as Nature put on its truly magnificent show. When it ended, we wondered if our next eclipse, on August 11, 1999, could possibly compete with what we had just seen.

Chapter Seven:
1999: The Voyage Begins.

A fter our successful voyage to darkness for the 1998 eclipse, catching the 1999 total eclipse from the north Atlantic Ocean near Nova Scotia seemed to be an odd plan. According to the statistics published by the Royal Astronomical Society of Canada's *Observer's Handbook*, the chances that the Nova Scotia sky would be completely clear were less than 20 percent! And the chances of seeing the eclipse through light cloud or fog was marginally better, at 30 percent. But having attended Acadia University for four years, I was quite familiar with the weather patterns in the area. In August and September, strong high pressure systems can park comfortably over Canada's maritime provinces for a week or more, bringing clear skies and warm temperatures. Of course, this time of year was also the start of the hurricane season, and already some hurricanes were threatening to march up the east coast. So as eclipse day grew closer, I tried to ignore those climate statistics.

Then came the rain of Sunday, August 8. As a large low pressure system hit the East Coast, our family boarded a limousine van for the drive to the docks at New York Harbor.

Despite the drizzle we were all excited. For the first time in years, this was to be a vacation for every member of Wendee's core family. Her parents, Leonard and Annette Wallach, were especially excited about the family time ahead. Annette was well known as the former director of Treasure Island Day Camp, a Long Island summer camp for several hundred children. Len, retired from a long career in dress manufacturing, has been losing his sight in recent years from glaucoma. Gail, a nutritionist, is Wendee's oldest sister. On the way to the docks she asked me if people in Connecticut would get a good view of the eclipse. "Why Connecticut?" I wondered as I explained that people in Connecticut, and in all the New England states, would see a deep partial eclipse with the Sun looking like a thin crescent. (Gail's new boyfriend Marc, I learned, lived in Connecticut.)

Joan-ellen, a registered nurse, is Wendee's other sister. It's fun to see Wendee and her sisters interact-the closeness among the three is a joy to see. Even Sandy, Wendee's younger brother, commented on how good it was to have everyone together again as we boarded the 23,000 ton *Regal Empress* around two o'clock that afternoon.

THE *OLYMPIA'S* VOYAGE TO DARKNESS

"The Voyage of the Cruise Ship Olympia," Leif Robinson explains, was "the first eclipse mass transit system." *Olympia* was a former name of the *Empress* which we were boarding at that moment. In 1972, this ship, even then an old, converted troop ship, became the first cruise vessel to sail pur-

The Regal Empress moored off St. Anthony, Newfoundland, August 15, 1999. Photo by Roy Bishop.

posefully into the path of a total eclipse of the Sun.[1] It sailed out of New York Harbor, as it did for us, but on July 8, 1972. On eclipse morning the passengers saw a clear sky but clouds were on every horizon. It seemed that *Olympia* was in the center of a hole in the clouds, and if she wanted her passengers to see the eclipse that afternoon, her captain, John Katsikis, would need to steer her southeastward along the path of totality. With only one small cloud to threaten the clarity of

1. Leif Robinson, interview, 29 April, 2000.

the sky at totality, the passengers set up their instruments and waited. The cloud faded away in plenty of time. "Several seconds before the first diamond ring flashed into view," a *Sky and Telescope* article described, "coronal streamers could be seen protruding from the east limb of the Sun, if the observer blocked out the remaining bit of photosphere. Then the hush that had fallen over the ship was broken by the insectlike clicking from hundreds of cameras."[2]

Olympia's voyage was the first major cruise specifically to see a total eclipse of the Sun. Phil Sigler and Ted Pedas, two college and university teachers, thought up the plan and saw it through. This first trip was a harbinger: By 1998 some 20,000 people saw the February 26 eclipse aboard cruise ships in the Carribbean, and about as many headed out to see the eclipse of 1999. All but one of the 1999 ships were headed to the Black Sea, which promised a clear, hot sky. Only *Regal Empress*, the former *Olympia*, tried to catch the eclipse from the North Atlantic. The two sites on the Atlantic—1972 and 1999—were not far apart—6 degrees apart in longitude, and 2 degrees in latitude.[3] In 1972 the Sun was 24 degrees above the horizon during totality; this year it would be only 3½ degrees. Some years later, the ship was chartered to see a second eclipse. We were there for its third visit. Now owned by Regal Cruises, the *Regal Empress*, the ship that started it all, was headed for yet another rendezvous with totality.

Just as her 1972 cruise was unique for its time, the

2. Edward M. Brooks, George S. Mumford, and Leif J. Robinson, "The Olympia's Voyage to Darkness." *Sky and Telescope* 44 (1972) 154–157.
3. 1972 position: 40.3N, 54.5W; 1999 position: 42.1N, +60.8.

Eclipse viewers are transfixed by the unfolding eclipse, on the Regal Empress, *five minutes after sunrise, August 11, 1999. Photo by Roy Bishop.*

Empress's 1999 cruise was unique—a trip out to sea for an eclipse at sunrise. The venture was risky—by far the majority of eclipse chasers had invested their vacation funds in cruises to the Black Sea, where the weather prospects were excellent and where the eclipse would take place with the Sun almost overhead. The *Empress*'s plan was to be in the path of the Moon's shadow less than a minute after it touched the Earth— and almost precisely at the spot where she watched the 1972 eclipse—and coincidentally at the spot through which the 1963 eclipse had passed. In 1972 the eclipse took place late in a summer afternoon with the Sun high in the western sky. This time, even if the sky were technically clear, low haze and fog could easily block our view of the eclipse.

MEETING WITH THE CAPTAIN

To assess our chances and to position the ship in the best possible place, the leaders of our group met with Captain Peter Schaab on the first night of the cruise. I thought the meeting had an ominous beginning as I wandered from our Cabin, No. 45 on "U" deck, through obscure passageways, up and down narrow flights of stairs, and finally, having failed to find the meeting place, made it back to cabin U45. With rain pelting down on the decks, I was most discouraged, and if Wendee hadn't urged me to try again, I would have missed the meeting altogether.

I'm glad I didn't. Captain Schaab is a most experienced sea captain, and we were impressed with his knowledge of the eclipse path and how to navigate the ship into it. The rest of us were all experienced eclipse chasers. Weather forecaster Joe Rao, amateur astronomer Sam Storch, and Ann Burgess, leader of Northstar Cruises who had organized the tour, sat with the captain and the ship's senior staff.

Some eclipse tours have had to put up with ship's officers with little knowledge or interest in natural phenomena like eclipses. With Captain Schaab, we didn't have to worry. I didn't even need to show him my small map that traced the path the Moon's shadow would follow; he knew where the shadow would fall to the nearest hundredth of a nautical mile, and he proposed a route that would get the ship safely within the narrow band of totality several hours before the event was to start-so long as everything went well. Our schedule had us staying in Halifax until 4:30 P.M. the day of the eclipse, and although we were concerned that we wouldn't make it to the shadow path in time, Schaab assured us that we would.

A CLOUDY NIGHT

Waiting for me after the meeting were five close friends who were joining us on the cruise. Leo Enright and Denise Sabatini, two of Canada's best known amateur astronomers, had joined the trip from their home on beautiful Sharbot Lake, north of Kingston, Ontario. Roy and Gertrude Bishop joined us from Avonport, Nova Scotia. And Patsy Tombaugh, widow of Clyde, the discoverer of Pluto, was with us from Las Cruces, New Mexico. Roy had been so worried about the ship's making it to the path of totality in time that he almost didn't come on the cruise at all. When I told him of the Captain's plan, he felt reassured about the ship's course but not about the weather forecast We went out on the ship's upper deck to watch the storm. There was a little rain, lots of wind, and occasional stars. The disturbance we were in was supposed to move out by morning, but a second one was coming in a day or so behind it. If the new storm moved according to forecast models, it would hit us shortly after the eclipse. We could still be clouded out.

As Roy and I stood looking up at the clouds racing by, we thought of other eclipses. I reminded him of the 1970 eclipse, where he was smart enough to travel differently from me, and he saw the total eclipse and I saw the darkened underside of a cloud. Two years later, the year of the Empress's first voyage, he saw a total eclipse from Nova Scotia and I saw a partial from Montreal. Here and now, we were trying again-a total eclipse visible from southeast of Nova Scotia. Whatever happened this time—cloud or clear—we'd go through together.

Chapter Eight:
Sailing closer

A s *Regal Empress* steamed toward Halifax through rough seas and stormy weather, Wendee and I thought about how we'd be the first to see the eclipse from the Earth's surface; less than a minute after the Moon's shadow first struck the Earth, it would pass over our ship. In fact, we would see the eclipse so early that when totality began for us, the entire shadow of the Moon would not yet have struck the Earth! At the onset of totality, the other end of the shadow would still be in space! I told Wendee how, after it passed us, the shadow would race across the Atlantic to Cornwall, England, and then travel near Paris, Munich, and on to Rumania, the Black Sea, Turkey, Iran, and Iraq. The shadow would leave the Earth over India, its whole crossing taking about three hours. While our ship would experience just 52 seconds of totality, the people in Europe would stand under the Moon's shadow for more than twice as long.

HOW TO OBSERVE THE SUN SAFELY
On this day at sea, the tour leaders gave presentations on

these issues. It was literally "astronomy day" on board, with informational sessions about how eclipses take place and how to safely observe them.

Eclipses can cause permanent damage to human eyes—most people know that, but few know why. Wendee told me of a fifth grader in her former school who insisted on staring directly at the Sun during the 1984 partial eclipse despite his friend's efforts to stop him. He lost his sight for three days afterward, she said, and was very fortunate to regain it after that. Frankly, I'm not sure how the foolish boy did regain his sight, and how it will be affected in the future.

The truth: Looking directly at the Sun is always dangerous. Because our Sun is a G-type star, a considerable portion of its radiation is ultraviolet. Our atmosphere's ozone layer absorbs most of this radiation, but harmful amounts of it do break through and reach us. We wear protective clothing or sun blocking creams because of this spectroscopic nature of the Sun. "We have the wrong star," astronomer Clyde Tombaugh once told me. "If we were orbiting a K-type star, we wouldn't need an ozone layer, and we wouldn't have to worry so much about skin cancers."[1]

What does all this have to do with the human eye? Ultraviolet radiation can be catastrophic for the eye. If you look at the Sun long enough, the UV radiation can actually burn a hole in your retina, an event that would result in permanent, though partial, blindness. Most of the time, the Sun's brightness prevents your eyes from remaining open in direct view

1. Clyde Tombaugh, interview, July 10, 1996.

for more than a few seconds. During an eclipse, the normal glare of the Sun is reduced to the point that you can look at it for long periods of time. However, the remaining visible portion still is sending out harmful UV rays. Without your eye's built-in protection system at work, you can look at the Sun longer, and therefore your eyes have a far greater chance of suffering UV damage.

To safely view the eclipse, we suggested the following:

1) Use a No. 14 Welder's Glass, or the special eclipse safety glasses that the ship provided to everyone on board.

2) With a small telescope, binoculars, or opera glass, project the Sun onto a white cardboard or a piece of paper. Do not look through an unprotected telescope, and do not use a telescope equipped with a filter at the eyepiece end. These filters could easily shatter due to the concentrated heat of the Sun.

3) Put a tiny pinhole through a piece of paper, and project the Sun's image through the pinhole to another piece of paper. But do not look at the sun through the pinhole.

THE STARS HUNG LIKE BASEBALLS

As the vessel steamed northeastward, the storm moved off and the sky began to clear. The Sun sank in the northwest, giving way to an evening glowing with stars. On the bow of the ship that night dozens of people gathered for our first star party. The sky was so dark and clear that we could taste the beauty of the eclipse that would come with the rising Sun a day and a half in the future.

"The stars hung like baseballs," my grandmother loved to

say about seeing the sky from the middle of the Atlantic Ocean. On this night, the Summer Triangle hung overhead, and the bright semicircle of the Northern Crown hung high in the west. For the first half hour, we pointed out constellations, joining make-believe classical Greek figure lines in the black sky. As the group thinned, Wendee and I noticed Roy Bishop off by himself, standing against the forward rail, peering off into space through a pair of binoculars.

"Take a look through these binoculars," he said, "and when you find something you're interested in, press this button." I pointed the binocs toward the Andromeda galaxy, a dim misty spot some two million light years away. After a few seconds of searching, the galaxy swam into the field of view. I tried to steady my hold on the instrument, leaning heavily on the rail.

"Now, push that button!" Roy repeated. Suddenly, the stars in the binocular field stopped racing about like fireflies, and the galaxy looked as steady as if I had mounted the binoculars on a firm tripod. "Wow!" I exclaimed as I discovered the magic of image-stabilized binoculars. It's an optical technology that allows lenses to act as if they're floating. Here we are two days from the eclipse, and already I have learned something I hadn't known before. The technology really isn't that new, actually. I first saw something similar to it during an Academy Awards presentation, when someone climbed a flight of stairs holding a camera, and its images were rock-steady. In any event, as Wendee and I excitedly headed back toward Cabin U45, we wondered what the next day would bring, and if a continuing clear sky would allow more playing with fancy binoculars.

Chapter Nine:
Horsetail Cirrus!

A t 5:30 AM, I awoke and looked toward our window, which was positioned in Cabin U45 as if it had been added as an afterthought. Carefully navigating the route from our beds to the window. I found the small curtain hung low over a small window. Somehow I didn't trip or bump my head in the soft, dawn light. I opened the drape to see a calm, blue ocean, a brightening dawn, and a brilliant blue sky. It was gorgeous! "Wendee," I explained, "I've got to run up and see the sunrise from the upper deck!"

I wasn't surprised to see Roy there, carefully scanning the horizon with his stabilizing binoculars and noting the ship's position with his global positioning system. "Oh!" he said on seeing me, "would that we have this kind of a morning tomorrow!" There wasn't a cloud in the entire sky, which was actually something to be concerned about, because fair weather in this part of the world usually means some sort of cloud cover—often a completely clear sky meant that a change in weather was coming.

As the sky continued to brighten, Captain Schaab appeared on deck. The man seemed to bring good luck with him; just as he appeared, we saw a splash in the distance, off to starboard. A breaching whale! "We'll see a lot more of those as we head north to Newfoundland after the eclipse," the captain said. Being on that silent ship, alone at dawn on the sea, with just a whale for company, was magnificent. Add to all this, a rapid brightening unfolded in the sky to the east, and then a flash of green light as the first rays of Sunrise shot into the sky. We watched silently for a few moments, and then Roy said, "if we see that tomorrow, it'll be just a thin crescent Sun rising."

That morning at breakfast, we heard that newspapers around the world were reporting preparations for tomorrow's eclipse. A strong low-pressure system was heading straight for England, where thousands were gathering for the eclipse. France and Germany were in the path of an earlier storm. After breakfast we all walked on deck to watch the vessel enter Halifax harbor, site of a great explosion in 1918. Today, though, the harbor was calm, beautiful, and warm, and the sky above it was clear, with cirrus clouds coming in from the northwest. Clouds? What's this about clouds? Meteorologist Joe Rao walked past, nervously looking up at those encroaching clouds. "Those clouds are not a good sign," he said.

Roy looked up, and shook his head. "Horsetail cirrus!" the man from Nova Scotia said emphatically. "See the shapes of those clouds, like horses' tails? At least in the Maritimes, that's a strong sign of fair weather ahead!" I would place my bet with Roy any day. Besides being familiar with the physics of climate and weather patterns, this native Nova Scotian has spent most of his life in the hills, valleys, and tidal basins of Nova Scotia. You can't do that in this rural province and not know its weather patterns as well as you know the rooms of your house.

So the *Regal Empress* pulled into Halifax, cut her engines, and docked. As a lone bagpiper played songs of welcome for us, our whole family prepared to disembark—Gail, Joan-ellen, Wendee, Sandy, Mom, and Dad in his wheelchair and me. Our goal: a drug store somewhere in downtown Halifax, for this was to be our "great Florine F adventure." In recent years, Dad has suffered from a rare affliction known as orthostatic hypertension, in which his blood pressure can fall precipitously and without warning while he does any aerobic activity no matter how mild, and then rise dramatically when he'd be lying down. He takes a drug called Florine F that controls the symptoms to some degree, but somehow, when he arrived on board the *Regal Empress*, he found he only had two day's supply of this essential drug. Happily, my beloved niece, Dr. Alison Stein, saved the day. After a phone call from dockside to give her the precise dosage and other particulars of the medication, we walked up and down the hills in downtown Halifax till we reached the drugstore. One we had the precious pills, we continued on to a historic park in view of Halifax's Citadel. Wendee noted that the sight of all seven of us walking up and down the hills of Halifax, pushing Dad, who was holding all the purses and coats in his wheelchair, must have been hilarious. Even Dad enjoyed the excursion.

On this magnificent day it looked as though the whole city was on holiday enjoying the weather. I recalled an earlier visit to this town, in December 1968—I was walking through downtown in a blinding snowstorm, with rare winter lightning. Hoping to make it to the train station in time to take the afternoon Dayliner back to my residence at Acadia University, I trudged through the snow, entered the big old building and approached the ticket counter. "Can I exchange my ticket for the evening train for the earlier one?" I inquired.

"Certainly," the agent said helpfully, then he stared at me. Tired from my walk through the snow, I stared back blankly. "Well, he added, can I have your ticket?" I looked through one pocket, then the others. I don't think I had more than thirty cents in them, far less than the amount needed to replace the ticket—and there was no ticket. With pockets inside out and a forlorn expression, I admitted that I must have misplaced the ticket.

"There's nothing I can do, sir!" The ticket agent said, and I wondered if I had enough change to even call someone. I stared blankly at him again, and then a stranger came up to the ticket counter. "I found this ticket to Wolfville on the floor," he said. "Would you like it for two bucks?"

"That's my ticket!" I exclaimed, "and you can have my thirty-five cents!" Half an hour later the single-car train departed the station, snowplow in front, and with each turn of the wheels I—minus my thirty-five cents—moved farther from my embarassing trip to Halifax and closer to my Acadia home. I resolved that day never to go to Halifax again. So, here I was, with my wife and family, back in Halifax on our great Florine F adventure, the day before the eclipse. We wheeled Dad up and down the same hills, now in reverse order, climbed the steps and reboarded our ship. Just before 5, PM, the *Regal Empress* pulled away from dockside with three blasts of her whistle, and began her trek out of the harbor. She passed the breakwater at the harbor's mouth, then turned to the southeast. Our family, along with Roy, Leo and Denise, Patsy, and a new friend, Kandra Kargo, were sitting at the stern watching the ship head closer to our destination: a patch of water at 60 degrees 46 minutes west longitude, and 42 degrees 7 minutes north latitude.

Onboard, we relaxed and talked about the next day's eclipse. "There's Joe on his knees," Wendee remarked as Joe Rao came by, map in hand, "praying to the eclipse gods that we have clear weather tomorrow morning!"

"There's this low pressure system that's going to keep the northeast U.S. from viewing the partial," Joe answered, launching himself into weatherman mode. "We remain under a little ridge of high pressure, a ridge in the upper atmosphere that should keep all but the high clouds out of the way. We're damned lucky!"

Indeed, we really seemed lucky that afternoon as *Regal Empress* steamed out of the harbor. We were right on time. For about forty-five minutes we headed out of Halifax harbor, then we turned southeast, then south, then southwest. As we looked on in increasing horror, it was 1968 all over again as our boat, our ticket to totality for which we had no replacement, no alternative, was *heading back to Halifax!*

Chapter Ten:
A Voyage to Darkness

Farewell to Nova Scotia, the seabound coast
Let your mountains dark and dreary be
For when I am gone away on the briny ocean toss'd
Will you ever heave a sigh and a wish for me

The Sun was setting in the west . . .

Nova Scotia folk song

As the *Empress* steamed back toward Halifax, we felt that the phenomenal weather we were enjoying wasn't going to amount to much if the ship couldn't get out of Nova Scotia's capital. Ann Burgess of North Star cruises, our cruise organizer, worried about a worst-case scenario, that one of the engines might be on fire. The ship made its turn so wide that most of the passengers seemed unaware of the course change, though half the passengers, for whom tomorrow's eclipse would interrupt their morning run or bingo, wouldn't have cared. Barring unforseen problems, we were scheduled to steam into the path by 4 A.M. on eclipse day. If the sky were cloudy there, we would have two hours to move along the track in search of a break in the clouds. I saw Roy walk up and back along the length of the deck, looking at the shore and at his Global Positioning System unit. "Tell me your GPS doesn't show us heading in wrong direction!" "I quipped I hope that whatever is happening isn't serious," Roy said, obviously dismayed. "At this rate, we'll miss totality!"

Now I need to emphasize the energy and single-mindedness with which eclipse-chasers go after their quarry. They study maps and schedules, and make arrangements with shipping companies sometimes years in advance. In planning for the February 1998 eclipse in the Caribbean, Roy had actually tried out a site on Montserrat Island three years earlier. Then booked the hotel and travel for a group of some thirty enthusiasts, including Wendee and me. When Montserrat began erupting a year later, there was still plenty of time for Roy and his group to make different plans. We were saddened by the event, and in 1997 Wendee and I, then on our honeymoon, even flew over the erupting volcano. But our new plan for 1998 was already in place—we were to see the eclipse from near Aruba aboard *Dawn Princess*.

For the eclipse that would come with tomorrow's dawn, we knew our chances of seeing it were not great from the cloud-ridden North Atlantic. Thousands of amateur astronomers did not even consider this site, opting instead for the better chances on the Black Sea, where skies were usually clear. In fact, Ann had recently seen the large group she had organized leave for Athens and their Black Sea cruise. And so here we were, peering at a beautiful sky, a ten-hour sail to the path of totality, and our ship heading in the *opposite* direction.

I decided to find out why. I walked inside, where I met Clodagh O'Connor, the *Empress*'s entertainment director. She smiled at me as though absolutely nothing was wrong, but cruise directors are good at that—on one cruise when the ship stopped at a rainy port, the cruise director told us to enjoy the liquid sunshine! So I said, "Clodagh,""why are we returning to Halifax? Is there a problem with the ship?"

"Oh, no!" she assured me. "We're going back to fetch two passengers who missed the boat." The captain didn't want them to miss the eclipse.

I was torn between feeling relieved and annoyed. Vastly relieved that *Regal Empress* was healthy, but wondering about the wisdom of a decision that would all but wipe out our window. The captain, it turned out, made his unusual decision because it was thought that the two tardy passengers were part of our eclipse group. I rushed back out to my waiting friends and told them the news that the ship was heading partway back to Halifax, where she'd rendezvous with a small blue boat carrying the two passengers.

"How can they do this?" Roy was aghast. "He's going to have to push his engines to make it!"

Ever the optimist, I argued that since we didn't have to return all the way to Halifax, we should be heading southeast soon. "It's going to be okay," I tried to assure everyone. "This will only delay us a bit. It could have been much worse." The captain then made his announcement that we were returning to pick up two passengers.

Wendee walked toward the rail, and soon we were all

watching a tiny blue boat emerge from the crowd of sailboats and motor cruisers that were roaming about the harbor. As the blue boat grew larger, the *Empress* slowed to a crawl and the small boat came alongside. The boat pushed up against us as we dropped some ropes and waited as the two passengers were lifted up one by one. "I'll bet those two will be too embarassed to see anyone once they're aboard!" Wendee said as the first passenger was whisked away. "If I were the passenger on that little blue boat," Roy added, "I would be too embarrassed to come aboard. I would fly to St. John's [our next stop after the eclipse] and then sneak back on!" A few minutes after the ropes lifted the second passenger, the two vessels separated. "Our family is together again," the captain announced as he turned the ship in a second wide arc back toward the southeast, and soon we were heading, once again, out of Halifax harbor.

What a relief! Having effectively lost one hour, we were glad that it wasn't an engine problem. At long last we were finally on our way. But as we steamed out of the harbor to the spot we had passed before, Roy looked at his watch, then toward Wendee and me, and shook his head. We knew that those passengers had cost us almost every minute of our two-hour window. We had to hurry if we were to make our rendezvous with a major cosmic event, now just twelve hours away.

Quiet, friendly conversation lasted late into that evening. "I saw an eclipse when I was a child,." Wendee's mother said. "In New York, around 1925."

"The edge of totality was at about 89th Street," Roy

asked, "wasn't it?"

"Yes," Mother agreed. "It got very dark and windy. I remember wondering if the eclipse would change the way the world is."

"Did you cheer when the Sun came back?" Joanie asked.

"No, I was seven. They said that if you didn't cover your eyes you'd go blind."

Patsy joined us later at the ship's stern as the Sun set in the northwest. We could still see the distant coastline of Nova Scotia. This was the second eclipse cruise on which Patsy joined us; we had also watched the 1998 eclipse with her. Doing astronomical things is bittersweet for her, since they remind her of the long and happy marriage she enjoyed with Clyde, a man we knew well and for whom, we understood, astronomy was *everything*. The sky, we noted, was quite clear save for some cirrus low in the west and north. We weren't too worried. Joe Rao's weather charts us still had us firmly in the middle of a weak high pressure system.

As the stars began to come out that night, somehow I remembered a night-before-the-eclipse long ago, back on July 19, 1963. I telephoned the recorded weather service to hear these ominous words, spoken with a strong French accent: "Chances for viewing eclipse in southwestern Québec, poor." I didn't expect to see the eclipse the following day. A fortuitous break in the clouds allowed me, my parents and my friend Paul Astrof to get a beautiful view of the sixty seconds of totality. Tomorrow, if all went well, we would see this eclipse again. The weather forecast was great, and if those propellers keep twirling, we should make it!

Northern summer twilights seem to go on forever. On this night the stars appeared lazily, one by one, as the Sun seemed to lie in waiting below the northwestern horizon. It was almost two hours before the faint arch of the Summer Milky Way unveiled its full majesty across the sky. What a marvellous sight! The small "circle citadel" of Corona Borealis, hanging like a cup high in the west, reminded me of what Gerard Manley Hopkins, the great English poet, wrote in February 1877.

> *Look at the stars! Look, look up at the skies!*
> *O look at all the firefolk sitting in the air!*
> *The bright boroughs, the circle-citadels there!*[1]

"Wow!" someone said as a meteor punctuated the night. With just two nights to go before the maximum of the Perseid meteor shower, this shooting star reminded us that the eclipse was not the only show at sea. The bright glowing trail of the meteor was caused by a tiny speck of dust crashing into the Earth's upper atmosphere, and heating the surrounding air so much that it glows. The dust was a part of a swarm that the Earth encounters every year as it orbits the Sun, debris from Comet Swift-Tuttle, discovered in 1862 and seen again 130 years later in 1992. This debris spreads around the comet's orbit, and every year, when the Earth crosses that orbit, the dust enters the Earth's atmosphere, heats up to incandes-

1. Gerard Manley Hopkins, "The Starlight Night," Gerard Manley Hopkins, *The Poetical Works of Gerard Manley Hopkins*, ed. Norman H. Mackenzie (Oxford: Clarendon Press, 1990), 139.

cence, and we observe meteor showers.

As the hour grew later, Wendee and I headed to the ship's stern for a midnight snack. Ann Burgess was there, also unable to sleep the night before an eclipse. Ann regaled us with stories of other eclipse cruises she had organized, including one with a Captain who insisted on turning the ship around and around *during the minutes of totality!* He did not understand that such a plan worked against anyone trying to photograph the eclipse, let alone see it comfortably. But she had high praise for Captain Shaab as one of the best she had ever worked with.

Then we returned to our cabin. Although I wanted to observe all night, Wendee suggested that I try at least to sleep for a couple of hours. We lay down in bed and talked a while. But I couldn't sleep. I had seen eclipses, but never one at sunrise on the ocean. At 3:45 I quietly left the cabin, and headed to the forward promenade deck.

Chapter Eleven:
Alone on Two Wide, Wide Seas

Alone, alone, all all alone,
Alone on a wide, wide sea
                    —Samuel Taylor Coleridge,
        The Rime of the Ancient Mariner, 1797–1978

W hen I arrived at the ship's forward promenade
deck at 4 AM, the sky was clear and sparkling.
Standing out near the bow, my orange telescope (which we called Pumpkin) in hand, I saw the ocean
stretching out toward the horizon. Corona Borealis, high in
the sky just a few hours ago, now sat low in the west. Toward
the north, another vessel's lights shone in the distance. As it
sailed northward, after half an hour its lights disappeared one
by one as it dipped below the horizon to prove, once again,
that the Earth is round. There were no other ships, and no
planes above—just the moving ship, the sea, and the sky.

Suddenly a faint flash of light brought me back to Earth.
Startled, I wheeled around and looked up. Two decks above
me, a bridge officer was lighting his cigarette. Enclosed by
the windows surrounding the darkened bridge, the officer,

responsible for directing the vessel's speed and bearing, might as well have been in a different universe—that's how dark and quiet it was on that deck, and how alone I felt.

It was time to get busy. Standing there with the telescope, I felt like an old sea captain myself. Pumpkin, however, was a lot more powerful than the small refractors that they peered through to sight icebergs, land, and distant ships. But Pumpkin had no interest in those earthly things. For the next hour, the objects I was searching for were not icebergs at sea but those in the sky: comets.

ICEBERGS IN THE SKY

Suddenly realizing this relationship between comets and the sea, I felt that a search for comets from aboard ship was a totally logical thing to do. Comets actually consist of a combination of ices (not just water ice), and rock. Icebergs depart from the great ice fields in the far north and south, and then drift through the oceans, occasionally endangering passing ships, as the passengers of *Titanic* found to their horror in 1912. Comets begin their journeys by leaving one of two large comet fields, the Kuiper Belt where Pluto lies, or the Oort cloud, much farther away. I thought of Patsy now sleeping several decks down, and how her husband Clyde had understood that the icy planet he'd discovered, Pluto, would have been "a comet for the ages" had it changed in its distant orbit and flown near the Sun. As comets leave the space docks of their births, they drift through the ocean of space where they can occasionally endanger passing planets, as the dinosaurs found to their horror 65 million years ago.

This may seem a cute analogy, but for me, on that morning, it was quite real. As I stood on the bow of the ship, I stared out into two oceans, the Atlantic and the sky. Both stretched out as far as I could see, and the ocean was so calm I could see stars and constellations reflected in it: in a sense, I couldn't even tell them apart. With this feeling of unity, calm, and utter aloneness, I pointed my telescope toward the northeast and began to search for icebergs in the sky.

A COMET SEARCH

Searching for comets is a happy pastime that has occupied many of my clear evenings and mornings since I began the program on December 17, 1965. It's like being a night watchman. During the more than 2,500 hours I have spent with my eye at the eyepiece, moving a telescope up, down, then up again, I have discovered 8 new comets. And through a photographic search program, Carolyn and Gene Shoemaker and I discovered another 13 comets, including the catastrophic Shoemaker-Levy 9 that collided with Jupiter in 1994. The quiet search hours have been both productive and pleasant. Even if I do not find a new comet on any given night—and that's true for all but eight of the thousands of nights I've spent looking—at the end of a night's search I invariably have a good feeling of accomplishment; I have searched a certain area of sky and declare it free of comets, or at least comets bright enough to be captured through my telescope. Thus each night's watching is completed successfully.

Because comets tend to be brightest when they are close to

the Sun, I usually search the region of sky that is closest to where the Sun has set or where it will rise, and so, on this morning, I moved the telescope back and forth in horizontal sweeps across the northeastern sky. As we raced toward our rendezvous with the Moon's shadow in just two hours, I expected that the vessel's pitch and roll would make viewing through a telescope impossible. But tonight's view was pristine: The sea was calm and the ship's motion resulted in only mild shifts of no more than a third of the field of view in the telescope's position as I swept slowly acrosss the sky. Even better, northwesterly winds were following us, so there was virtually no wind to bother me up at the ship's bow. Since looking for comets is an activity that involves absolutely no mental concentration, as my eye focused on what the telescope was showing me, I was free to contemplate the darkness of the night and the sublime meeting of two magnificent oceans.

Then I moved the telescope over one more field, and my brain slowly shifted its concentration to the faint fuzzy object that suddenly was staring at me. This region of the sky, near Auriga the Charioteer, is famous for its several clusters of stars. But the fuzzy object I was looking at was totally unfamiliar to me. For a moment the unlikely was a real possibility—just two hours before a total eclipse of the Sun, could I actually have discovered a new comet?

As it turned out, I *was* looking at a comet, my first iceberg sighted on this voyage. But it wasn't a new one. I later identified it as Comet Lee, a comet that had been in the evening sky and which had recently moved far enough east to enter the morning sky. The comet incident added to the

Sun setting behind the telescopes of Kitt Peak National Observatory. Photo by David H. Levy.

unique feeling of utter alone-ness—not loneliness, but a feeling, indeed, of being alone on two infinite seas.

Then, near the horizon in the east, stars were disappearing. As the sky brightened with the onset of dawn, I could see why—clouds were starting to build in the direction we were headed.

Chapter Twelve:
Eclipse!

With the onset of dawn, my private reverie came to an end. Roy was the first to appear. He wanted to come out to see the zodiacal light, and it was there, a faint tepee-shaped arc of light leaning at a sharp angle from the northeast horizon almost half way to the zenith. Then meteorologist Joe Rao appeared with his friend Sam Storch, an avid astronomy teacher who emphasizes the common aims of science and philosophy. A deeply religious man, Sam tries to blend science and philosophy, and even religion, in what he teaches his students. "There's nothing you can do to change an eclipse," Sam thought about the upcoming event. "It was ordained four billion years ago that this would happen, today, no matter what. I think it's an obligation to be here; you want to mesh with the mechanical gearing of the solar system."

Meanwhile, Joe had more earthly concerns. "Where'd those clouds come from?" he looked up and asked of the sky.

"We're within the path of totality," Roy volunteered as he examined his GPS. "Maybe we should move the ship away from the clouds." Joe left to confer with Captain Schaab. The clouds were not thick, but they seemed to be concentrated to the east, directly in the ship's path. And as the sky continued to brighten, they did not seem to get any higher. That was a good sign.

Up on the bridge, the captain noted the clouds also. "By 5:30 in the morning," he told me later, "I was apprehensive about them. I didn't know exactly how they were moving; whether they were moving further into the path or not." The single layer of stratocumulus was quite low and did not seem to be growing in size or moving toward us.

It was time to wake everyone up. Returning to our cabin, I sat down on the bed. "Is it time?" asked Wendee drowsily. "It is time," said I. As she awakened and prepared to join me topside, I went down one deck to make sure Patsy Tombaugh and her friend were ready. "It's mostly clear," I said cheerily, "and I think we're going to do some eclipsing today!"

Back in our cabin, Wendee was now ready for her day. We took the elevator up five decks and went to our site, high on the promenade deck not far from where I had been observing half an hour before. It soon became obvious, though, that our prime site wouldn't do. If Captain Shaab were to turn the *Empress* to the north, our site would not get a good view. So we wandered to starboard and chose a new site near the front end of the starboard sun deck.

The sky was brightening fast, and the *Regal Empress*'s sun deck was filling up with groups of travelers staking out their viewing sites.

Patsy reminded us that she had seen this eclipse before, two saros cycles ago, in 1963. But Clyde was so interested in having her time his experiments during the minute and a half of totality that she barely got to see anything. We hoped to correct that oversight this morning, as the results of our planning and hoping were already coming to bear—the partial phase had already started, and we waited to see it when the Sun rose in the east. That event was now just fifteen minutes away, and still the ship was cruising eastward. The passengers were getting restless, especially Roy, who had a bevy of cameras set up at a spot that was not pointing anywhere near the rising Sun. "David," he said, "are they going to turn or not?"

Frustrated, I jumped over the rope that said Crew Only and raced up the steps two at a time toward the bridge. There the other tour organizers had set up their spots, and with a full view of the sky all around, they did not understand the confusion the rest of us were feeling. I was told that the ship was doing a final positioning to get the Sun in between the two areas of distant clouds to the east.

What I didn't know at the time was that the captain was as anxious as any of us to stop the ship and position it, but Joe was still trying to keep the ship going further south, away from the clouds. Captain Schaab now felt that further fiddling was just plain unnecessary. "By ten past six, quarter past six," he told me, "I knew that the clouds were not moving. We had observed and measured them for a good hour, and I was confident that it was going to work out." Although this would not be his first eclipse, it was our captain's first as an adult, and he was as anxious as the rest of us to see it. "We viewed the first one when I was about five years old, with a small black glass burned over a flame," he confided later. "I saw the outline of the Sun and the black shadow moving over it."

Although Joe was still pushing for more maneuvering, Captain Schaab finally had enough. "Joe," he said decisively, "I have fiddled, and I have fiddled, and I will fiddle no longer. I am turning the ship to port!"

SUNRISE

I returned to our deck and told Roy and the others that the ship was about to turn to port. But all the while, something amazing was happening in the sky, which had been brightening rapidly with the approach of sunrise. Now the increase in sky brightness slowed down, then strangely stopped, and began to reverse-the *sky began to get darker just before sunrise*. Then a thin crescent Sun rose out of the sea, and as it cleared the horizon it just sat there, its bottom cusp resting on the waves like a sunbather dipping an idle toe in the pool. "The first special effect," Wendee would write after the event, "was the sight of the Sun rising in partial eclipse. The crescent Sun was phenomenal, and made this eclipse unique in itself."

When the Sun or the Moon is that low in the sky, it looks much, much bigger than it is when it has some altitude. It is called the "Moon illusion, caused the by fact that our minds have horizon markers to compare with the size of the distant Moon or Sun. When the orbs are high in the sky, we have no such basis of comparison and they seem smaller. On this morning, both Moon and Sun were locked in a giant Moon illusion, and so close to the sea they looked even larger.

As the crescent Sun rose higher, it rapidly got thinner. The ship was strangely quiet as we looked out to sea and tried

Sunrise at 42 degrees 10 minutes N, 60 degrees, 39 minutes W. 6:10 a.m., August 11, 1999. Photo by Roy Bishop.

to take in the whole unearthly scene. And it wasn't just the Sun. The *sky* was now darkening steadily—it was much darker than it had been a few minutes before sunrise. And with the darkening sky, the *sea* seemed to be quieting down. Even though the starboard side was now crowded with passengers, I felt just as I had a few hours earlier, alone on the ship, alone on a wide and darkening sea.

The Sun was now well over the distant low clouds, and the ship, her starboard side facing the Sun as planned, slowed down. After all the waiting and uncertainty, with less than a 20 percent chance of seeing this event in a perfect sky, it was now certain that we were going to see it in a perfect sky. Roy turned away from his camera and tripod, walked over, and we embraced. Wendee and I hugged each other too.

THE CRESCENT GETS THINNER

All the passengers were reacting to Nature's mood of quiet suspense. Conversation slowed down; people seemed to talk in shorter sentences. The event at sea had all our attention. I suggested that everyone watch the sky darken. With protective glasses on, we could watch the solar crescent shrink. Then, with glasses off, we watched the sky darken as fast as though someone had just thrown a celestial dimmer switch. Wendee's dad, with eyesight so poor he can no longer drive, was nonetheless able to make out the crescent Sun through his specially filtered glasses, and he readily noted the darkening sky and the shadow of the Moon racing towards us at 12,900 miles per hour (that's 215 miles per minute). Patsy was overwhelmed by the event and her surroundings. "This is such a quiet time," she said softly, as if we were all getting ready for the end of the Universe. We could easily see how ancient cultures would be terrified by a solar eclipse. Even with my own sense of understanding, something inside me wanted to reach out and stop the oncoming tide of darkness.

There were four minutes left to totality, and still the sky grew darker.

"Look how dark the water is," Wendee pointed out. The *Empress* was now heading northwest at a snail's pace, just

As the eclipse turns into a diamond ring, the shadow is about to leave the Sun and continue on its journey across the Atlantic to England. Photograph by Kandra Kargo.

After it left us, the shadow crossed the Atlantic and over Europe. In this European Meteosat 6 image from 22,000 miles up in space, the shadow races over Eastern Europe.

fast enough to let her water-filled stabilizers keep the vessel steady. The Atlantic was so still that we could see a hundred-thousand crescent Suns poetically reflected in the water. Roy and Wendee were snapping pictures of the dying cresent Sun, now risen to three degrees above the horizon. The rest of us just stared.

The darkness was coming in waves. I looked up to see a sky that was darker than it had been before sunrise. It was unnatural, so soon after sunrise, for the sky to be returning to night. We all seemed to realize that the eclipse, at this point, was far more than an embrace between Moon and Sun. The darkening of the sky was accelerating. The sea was still displaying its million thinning crescents. All around us there was an unearthly hush.

And still the sky grew darker.

Chapter Thirteen:
A Diamond in the Dark

With the crescent now just a thin, curved line, everything was still—the sky, the sea, even the crowd. Some could actually start to see the Moon's shadow dropping out of the sky to the west. "Look how dark it's getting!" Joan-ellen exclaimed.

Patsy watched the crescent Sun shrink from a curve, to a thin line. Then something startled her—a ripple of darkness rushed beneath her and raced out to sea. She had caught the rarest phenomenon of the eclipse, an effect called shadow bands. "I thought I was seeing things," she explained later. She was not. "The shadow went right under my glasses." The shadow bands occur seconds before or after totality as sunlight races through valleys at the very edge of the Moon. Edward Doyle, a passenger from Manhattan, also saw the shadow bands describing them as "a series of moving shadows, very faint, and very fast. They lasted less than a second, coming from the front to the back of the ship."

"On my mark," Joe Rao's voice crackled through the ship's loudspeaker. "Sixty seconds to totality!" I remembered my dad's words from 1963: "Please give my son a

b   r   e   a   k
in the clouds." This time his prayer would be answered
fully, for the sky was almost completely clear. Through the
special sun filters, the solar crescent was shrinking to barely
a spot. Just then Wendee removed her glasses to snap a picture of the Sun. She was about to replace them when she
glanced up—

"Oh David-David-DAVID! *Diamond ring!*"

I ripped off my glasses and looked. The crescent was now
just a bright point of sunlight, and the corona surrounding
the Sun was bursting into view. "*Glasses off everyone!*" I
yelled. I was trying to ensure that everyone took their filters
off, for during the total phase of an eclipse, when the Sun is
completely covered by the Moon, there is no danger in looking directly at the sight

"Ahead of schedule!" Joe Rao's understatement came
through the loudspeaker. The quiet on the ship ended in a
burst of excitement just as the Sun vanished.

In its place was a jewelled crown.

My telescope was patiently waiting; if it was to get used,
now was the time. I yanked its filter off and peered through,
then silently gave it to Wendee.

"Oh David! There's prominences *all over it*! Joanie, come
look!" Wendee explained that the prominences looked like a
ring of rubies. Joan-ellen put her eye to the scope and uttered
on involuntary "Oh *wow!*" Those ruby prominences were the
highlight of the view through the scope, she said later. We
could see them for the tongues of flame they were, each one

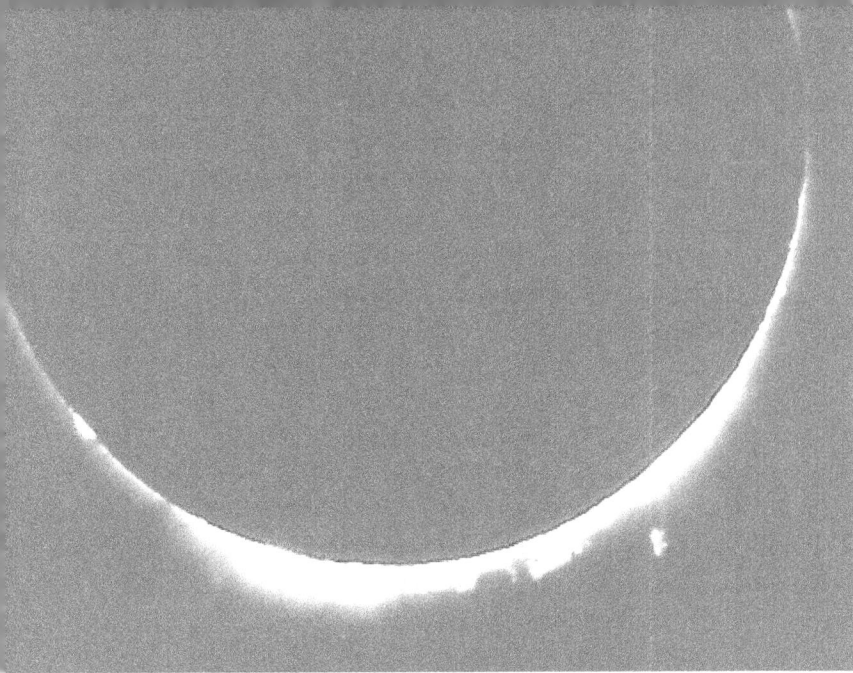

Prominences erupting from the southern edge of the Sun; August 11, 1999. Photograph by Greg Babcock.

arcing majestically away from the Sun, each one fully capable of swallowing the Earth. I've seen these before during eclipses but never so large or so ruby-red—their closeness to the horizon must have intensified their color.

While the rest of the group was looking through the telescope, I glanced toward the horizon again. The seconds were ticking by. The total eclipse utterly dominated, but it didn't complete the view. Surrounding the Sun, and extending more than halfway around the sky, was the cigar-shaped ribbon of darkness that was the shadow of the Moon, more clearly visible in the dark sky than I had ever seen it before. But it was moving fast—rocketing past us at more than twelve thousand miles per hour. Like a great hour hand, the shadow spun around to the northeast. Very little of it remained southwest of the Sun, and it was moving away fast. One of my fellow travelers noted how mystical, how quiet it was at totality: "I thought of the primitive people being absolutely terrified. I wasn't terrified but I had the feeling of people who used to be." I also noticed that looking out to sea, all was incredibly quiet. Millions of people would watch this eclipse as the shadow continued its journey, but not here, and not yet. There were no planes overhead and no vessels around us. No one. Just our ship, the sea, the sky, and the spectacle.

"I expected the diamond ring," Joan-ellen said later, "but I wasn't ready for the ruby prominences around the Moon. Also not prepared for how brilliant the Sun was around the Moon. It never got totally dark. I was not ready for that corona. I was so impressed. I do not want to buy a picture, because that's not what I saw. I will have the memory for-ever."

"This eclipse was completely different from what I was used to." added Denise, who has seen several other eclipses. "It just overwhelmed me. The fact that I was able to see the prominences with the naked eye!"

Despite his poor eyesight, Len Wallach had no trouble discerning the details of the corona. "It was very impressive," he said. "Overpowering." Standing next to him, Annette felt the chill of a cool breeze. Then there was a brightening near the top of the Sun. A second later, a spectacular shaft of sunlight burst past a mountain at the edge of the Moon. "A beam shot out," Annette recalled as another spectacular diamond ring marked the end of the total phase. As fast as it had begun, it was over. "I realized how little we are."

Up near the bridge, Sam Storch stared in silence as the diamond ring widened into a shaft of sunlight and then a thin crescent, and did what seemed the natural thing. With subdued emotion, he began to pray. "*Baruch atoh Adonai, elohenu melech ha-olam; osay ma-ah-say vareyshees.*" The translation: "Blesed art thou, Oh Lord our God, King of the Universe, who has created the wonders of heaven."

We stood there for a second, moved beyond words. Joanellen broke the silence with a simple statement of fact: "What–a–show."

As though a dam had burst, people started cheering, yelling, and hugging. On the bridge, Joe noted that of the eight total eclipses he has seen, "This is the only one during which no clouds were encroaching on the Sun."

"Shall I blow the whistle?" a crew member asked. And the response was an enthusiastic "Yes!" Thus, *Regal Empress* joined in the mayhem; standing alone in the Atlantic, three loud, long blasts sounded a salute to the Sun, in thankful farewell to the shadowy visitor from space that had visited us ever so briefly. "This eclipse went right to the marrow of my bones," said the captain. "The enthusiasm of this group—even before the eclipse—really got me going. With the Sun, the Moon, and a little bit of help from Joe Rao, I don't know how it could have worked out any better."

"This," Wendee emphasized to her whole family, "is what people chase. This is what people spend a lifetime chasing."

SECOND SUNRISE

"Ladies and Gentlemen," Joe intoned. "A little ahead of schedule, but nonetheless you've just seen the last total eclipse of the Sun in the millennium!" (Actually, it wasn't the eclipse that was ahead of schedule, but our ship, as Joe later calculated. We were positioned closer to the western edge of the path, the very beginning of the shadow's trek that day, than he had realized.)

I replaced the filter on the telescope, and let people look at the Cheshire Cat grin of a crescent that now was the Sun. And as time passed, other passengers shared their own views of what they had experienced. "This is the most striking eclipse I have seen!" Roy exulted. "The sky was clear, the Sun three degrees above the ocean, and there was a necklace of rubies. There was a less than 20 percent chance for what we had!" I agreed. With Wendee, her family and our friends with us watching the jewelled crown hanging over the sea was an experience I won't forget.

"Eclipses have personalities," said Ann Burgess. "The 1998 eclipse had a purplish quality to the sky, perhaps because totality was longer. This one seemed more joyful, partly because the Sun was just rising and we had the whole rest of the day after it."

"It was so emotional after totality," Wendee said. "All over the ship people were hugging each other, teary eyed. "Hours afterward," someone added, "when I thought of what I saw, I burst into tears."

Chapter Fourteen:
Here Comes the Sun!

'The moving finger writes and, having writ, moves on.' Like a moving finger of darkness the cone-shaped shadow of the moon had dipped down, scrawled its brief two minute mark of night across the land and then moved on, still writing, but now with invisible ink upon the empty page of space.

Leslie C. Peltier, Starlight Nights, *1965*.

Across miles of ocean and half a continent away, Clark and Lynda Chapman were quietly sleeping in their Rocky Mountain home near Boulder, Colorado. Clark is a planetary scientist who has spent the last several years studying images of moons that are more distant than ours. A member of the Galileo spacecraft's imaging team, he was looking forward to spending his day studying the latest spacecraft images of Jupiter's enigmatic and ice-covered moon Europa. As Clark awoke, he lazily turned on his radio, and was startled to hear my voice describing the eclipse I had seen moments earlier.

As the shadow of the Moon raced across the cloudy

Atlantic to its rendezvous with England, at our site the Sun was a crescent once again, but now facing the other direction. I looked at my watch and headed inside to our cabin to await that telephone call, via satellite, from National Public Radio. I returned to the empty cabin and sat on the bed, relieved for a couple of minutes of privacy. The cabin was brightening rapidly as the Sun's presence grew ever stronger, and then the telephone rang.

What a way, I thought, to begin a new day. I see a total eclipse of the Sun, and then I get to describe it to the world! And as Clark's radio program went on to the rest of the day's news, he was pleased that, in a world whose day was filled with a war in Indonesia and a shooting in California, that the radio station allowed its listeners to leave Earth for a few moments to join the moonshadow as it paid its precious visit to us.

The interview completed, I headed back to the appropriately named sun deck, a jumble of chairs and tables still in place for the eclipse. Now, however, there was some friction. A passenger who either didn't know or care about the eclipse was unhappy that the path of her morning walk was partially blocked by people staring at the Sun through strange glasses. The best part of the eclipse was over, but a million crescent Suns were still to be seen reflected in the ocean waves.

SHADOW MOVES ON . . .

Although we were probably the first to see the shadow from the surface of the Earth, we were not the first to see the eclipse. According to associate editor Dennis di Cicco of *Sky and Telescope* Magazine, the first expedition to see totality was probably the one led by John Hopper from Massachusetts. Flying at 41,000 feet in a three-engine Falcon 900B, his group actually saw over the edge of the Earth and witnessed ten seconds of totality just before the shadow struck the Atlantic. Dennis himself was part of the second group that watched from an airplane as the shadow swooped out of the sky. "The view from 25,000 was stunning," he wrote to me. "We had plotted a very careful course and programmed the aircraft's autopilot to fly directly down the centerline. I did not expect the view we got—horizon colors, prominences, and shadow, plus a Moon illusion enhanced view of the corona made the naked-eye view spectacular." Dennis was surprised that our view was as good as it was. "At the start of the track (a little west of us) it looked 100 percent cloudy below, but there were breaks below us even at the short distance up the track we flew. Actually, the column of sunlight reflected from the water in the moments before totality was stunning."[1]

Already the shadow was reaching Cornwall, England, where massive crowds were braving skies made doubly dark by heavy clouds and the eclipse. Near Paris, the eclipse had a special majesty, for it was a triumph of prediction by one of the great 19th-century French popularizers of astronomy, Camille Flammariom. Writing in his book *Popular Astronomy* in 1885, Flammarion noted that the track of the August 11, 1999 eclipse would pass right through Paris. The map was a little off—the eclipse tracked just outside the city—but it was

1. Dennis di Cicco to David H. Levy, 1 September, 1999.

close enough to be a consummation for a man who badly wanted to see a total eclipse over his beloved home.

I was especially worried about my friend Peter Jedicke, who was watching the eclipse with his family in Germany. Since Peter's view of the 1991 eclipse in Hawaii had been blocked by an unusual summer low pressure system, he really deserved to see this one. But the uncertainties of weather and climate aren't necessarily fair, and Peter was forced to endure heavy clouds for a second time. The shadow then crossed over the much clearer sky of the Black Sea, where thousands of people stood aboard ships all lined up to observe the eclipse.

As the eclipse advanced through Italy, Pope John Paul II enjoyed its partial phases, as did Egypt's President Hosni Mubarak from Cairo and Israeli Prime Minister Ehud Barak a short distance away in Israel. But the eclipse did bring out the worst in some people. Even though Brazil was nowhere near the path of the eclipse, according to one newspaper, a police-man in Picul released three prisoners because he allegedly thought the world was about to end. A television station in Bulgaria did not cover the eclipse at all, and later apologized to its viewers that its camera crew was delayed in an erotic film shoot. In addition to these bizarre events, there were at least two tragic ones: In Cairo, Abdul-Nasser Nuredeen was charged with killing his wife after she refused to make him a cup of tea, preferring instead to watch the eclipse. In Romania, where the eclipse was total, a 31-year-old mother killed her newborn baby because she feared that the eclipse had cursed it.[2]

2. These stories come from "News of the Weird: Eclipse Madness," a column in *Chicago Reader*, 17 October, 1999.

Back on the *Empress*, about a quarter-hour after totality, the captain turned the vessel slightly to the northeast, and picked up speed as it began its journey to St. John's, the capital of Newfoundland. All the rest of that day we were at sea, out of sight of land and other vessels as clouds from an approaching storm began to build. By the end of the day the sky was mostly cloudy. After dinner, we sat as usual near the stern, relaxing and going over the events of the day. As we talked in the dimming twilight, we noticed that the ship had slowed to a very leisurely pace. We didn't know it at the time, but a piston had broken in one of the two giant engines. The ship's engineer was miraculously able to install a new piston, but we were delayed several hours getting into St. Johns. That engine could easily have died a day sooner, and despite a clear sky and calm sea, we would have missed the total eclipse. As it turned out, we entered St. Johns' beautiful harbor at sunset, and enjoyed a clear sky and the maximum of the Perseid meteor shower with members of the St. John's Center of the Royal Astronomical Society of Canada.

A few nights later and accompanied by breaching whales, *Regal Empress* rounded the northeastern tip of Newfoundland and entered the Strait of Belle Isle, and a day later the Gulf of St. Lawrence. We were at our northernmost point. Wendee and her sisters Gail and Joan-ellen were hoping for a view of the Northern Lights. It was cloudy that night and the next day it rained, but evening brought a crystal clear sky. I walked back to my now-favorite spot near the

Empress's bow. Although it wasn't late, no one else was there and I had the sea and the sky to myself, just like that never-to-be-forgotten morning before the eclipse. But this night was different. We were no longer on the ocean. To the south lay the north coast of Newfoundland, and to the north I could see Quebec.

There was also a bright glow in the northern sky.

Aurora Borealis! During the eclipse, we saw plenty of sunspots crossing the Sun, so we knew that the Sun was active enough to produce a display of the northern lights. Just at that moment, Roy appeared. "Roy! We've got aurora!" We rushed to our cabins to get Wendee and Gertrude. "Well?" asked Wendee. "It's happening!" Passing the word on to her sisters, we rushed back up to the deck.

Over the next two hours the crowd thickened as word about the display spread. The ship's staff ensured that all forward lights were off to enhance our view—and what a view! We stood spellbound as the glow erupted into a beautiful arc with shimmering green and red searchlight rays climbing the sky. Beneath that arc was a smaller, more distant arc, also with gossamer rays pointing upward. The show was visible in the sky, and it was reflected in the sea.

It was not enough for the Sun to be the star of a total eclipse. On this night the Sun perfected our cruise, and an astronomical experience never to be forgotten, by producing its dazzling encore in a show of the northern lights. "At first the aurora simply looked like a thick cloud arcing across the sky," Wendee wrote in her diary. "Then it began to swell and

Voyage end; packing up to leave: l-r: Gail Zimmer, Denise Sabatini, Leo Enright, David and Wendee Levy, Michael and Lily Falk, and Gertrude Bishop. Photo by Roy Bishop.

light shafts formed above and below it until it had the illusion of being backlit by spotlights from below. We even saw a brief show of red flames." As we stared at the sky, we felt as though Nature was performing just for us.

Appendix A:
How to View an Eclipse Safely

1. Looking directly at the Sun is always dangerous. Its ultraviolet radiation can be catastrophic for the eye. If you look at the Sun long enough, UV can actually burn a hole in your retina, an event that would result in permanent, partial, blindness. Most of the time, the Sun's brightness prevents your eyes from remaining open, in direct view, for more than a few seconds. During an eclipse, the normal glare of the Sun is reduced to the point that you can look at it for long periods of time. However, the remaining visible portion still is sending out harmful UV rays. Without your eye's built-in protection system at work, you can look at the Sun longer, and therefore your eyes have a far greater chance of suffering UV damage.

2. Use a No. 14 Welder's Glass, or special eclipse safety glasses that are often available at camera and telescope stores before an eclipse.

Or

3. With a small telescope, binoculars, or opera glass, project the Sun onto a white cardboard or a piece of paper. Do not look through an unprotected telescope, and do not use a

telescope equipped with a filter at the eyepiece end. These filters could easily shatter due to the concentrated heat of the Sun.

Or

4. Put a tiny pinhole through a piece of paper, and project the Sun's image through the pinhole to another piece of paper. But do not look through the pinhole.

5. You can see a crecent Sun projected by spaces between leaves by looking at the ground in the shadow of a tree.

6. During the total phase of eclipse, when the Sun's

The correct way to view the partial phases of a solar eclipse, using No. 14 welder's glass. July 11, 1991, Santiago, Baja California. Photo by Roy Bishop.

corona and prominences are visible, but not the bright photosphere, it is completely safe to look directly at the Sun, even through a telescope. Protection is needed as soon as the photosphere reappears.

Appendix B:
Future Solar Eclipses

I n the next twenty years, eclipses of the Sun will cross a variety of paths over the world. Here is a list of what's in store: (Total eclipses are in bold:)

Remember: *Do not ever look at the Sun without proper protection for your eyes. Permanent blindness can result from even a quick look. Normally the Sun is so bright that your eyes are forced to squint, then quickly turn away, as a built-in protection. But during an eclipse, when the Sun is partly obscured by the Moon, you are tempted to look at it longer and more intensely. The Sun's ultraviolet rays can actually burn a hole in your retina, resulting in permanent, partial blindness. A Welder's glass (No. 14 strength), or specialized eclipse glasses that are available from telescope stores, will block enough of the Sun's Ultraviolet rays to make it safe to look through.*

During the total phase of a solar eclipse, when the Sun is completely covered by the Moon, it is completely safe to look at the Sun. Protection must be in force again, however, right after the end of totality.

Date:	Saros	Type	Description
December 25, 2000	122	Partial	A deep partial eclipse covering all of the continental United Stares and most of Canada.
June 21, 2001	127	Total	Southern Atlantic, Angola, Zambia, Zimbabwe, Mozambique, Madagascar
December 14, 2001	132	Annular	Crosses the Pacific Ocean and Central America
June 10, 2002	137	Annular	Also crosses the Pacific Ocean; partial in U.S.
December 4, 2002	142	Total	Starts off the coast of Africa and covers a small area already covered by the 2001 eclipse, including Angola and Zambia, then through Botswana, Zimbabwe, South Africa, Mozambique, then crosses the Indian Ocean to southern Australia.
May 31, 2003	147	Annular	Arctic; partial in Asia
November 23, 2003	152	Total	Antarctica
April 19, 2004	119	Partial	South Atlantic
October 14, 2004	124	Partial	Pacific
April 8, 2005	129	Annular/ Total	South Pacific and northern part of South America; Eclipse is total in eastern Pacific only.
October 3, 2005	134	Annular	Africa
March 29, 2006	139	Total	Eastern Brazil, Atlantic, Africa including Egypt
September 22, 2006	144	Annular	Southern Atlantic
March 19, 2007	149	Partial	Deep partial eclipse in Eastern US and Canada

Date:	Saros	Type	Description
September 11, 2007	154	Partial	South America, Atlantic
February 7, 2008	121	Annular	Antarctica
August 1, 2008	**126**	**Total**	**High Arctic in Canada, Greenland, Arctic Ocean, Russia, Mongolia, China**
January 26, 2009	131	Annular	South Pacific
July 22, 2009	**136**	**Total**	**Asia**
January 15, 2010	141	Annular	Indian Ocean
July 11, 2010	**146**	**Total**	**South Pacific, Easter Island, Chile, and Argentina.**
January 4, 2011	151	Partial	Africa
June 1, 2011	118	Partial	Northern part of North America
July 1, 2011	156	Partial	Africa, South Pacific
November 25, 2011	123	Partial	Antarctic region
May 20, 2012	128	Annular	Pacific Ocean, western U.S.
November 13, 2012	**133**	**Total**	**Northern Australia, South Pacific**
May 10, 2013	138	Annular	South Pacific
November 3, 2013	**143**	Annular/ Total	Atlantic, central Africa (total except for beginning and end of path.)
April 29, 2014	148	Annular	Just one place near south pole
October 23, 2014	153	Partial	Western North America
March 20, 2015	**120**	**Total**	**North Atlantic Ocean, North of Scandinavia**
September 13, 2015	125	Partial	Southern Indian Ocean
March 9, 2016	**130**	**Total**	**Western Pacific Ocean**
September 1, 2016	135	Annular	Africa, western Pacific

Date:	Saros	Type	Description
February 26, 2017	140	Annular	South America, Western Atlantic Ocean
August 21, 2017	145	Total	Pacific Ocean, Oregon, Idaho, Wyoming, Nebraska, Missouri, Illinois, Kentucky, Tennessee, North and South Carolina, Atlantic Ocean
February 15, 2018	150	Partial	Antarctica
August 11, 2018	155	Partial	Deep Partial in Europe
January 6, 2019	122	Partial	Deep Partial in the Pacific
July 2, 2019	127	Total	South Pacific, Chile, and Argentina.
December 26, 2019	132	Annular	Indian Ocean
June 21, 2020	137	Annular	Indian Ocean
December 14, 2020	142	Total	Pacific, Chile, Argentina, Atlantic

and, for good measure:

April 8, 2024	Total		Mexico, Texas, Oklahoma, Arkansas, Missouri, Kentucky, Illinois, Indiana, Ohio, Pennsylvania, New York, Vermont, New Hampshire, and Maine, New Brunswick, and Newfoundland.

Appendix C:
Future Lunar Eclipses

E clipses of the Moon take place less frequently than their solar counterparts, but since each one is visible over the entire hemisphere of the Earth over which the Moon is in the sky, they are more frequently visible.

Unlike Solar eclipses, lunar eclipses can do no harm to your eyes; they are completely safe to view through unaided eye, binoculars, or telescope. The dates are in Universal Time, so are correct at the longitude of Greenwich, England.

Universal Date:	Saros	Type	Description
January 9, 2001	134	Total	Asia, Africa, ending in east North America
July 5, 2001	139	Partial	Asia, Africa, Austalia, far west N. America
December 30, 2001	144	Penumbral	Over North America
May 26, 2002	111	Penumbral	Asia, Australia, North America
June 24, 2002	149	Penumbral	Shading too light to be detectable
Universal Date:	Saros	Type	Description

Universal Date:	Saros	Type	Description
November 20, 2002	116	Penumbral	North America, Africa, Europe, Asia
May 16, 2003	121	Total	North America
November 9, 2003	126	Total	North America, Europe, Asia, Africa
May 4, 2004	131	Total	Europe, Africa
October 28, 2004	136	Total	North America, Europe, Asia, Africa
April 24, 2005	141	Penumbral	Pacific, western North America, eastern Australia
October 17, 2005	146	Partial	Pacific, western North America
March 14, 2006	113	Penumbral	North America, Europe, Africa
September 7, 2006	118	Partial	Asia
March 3, 2007	123	Total	Africa, North America, Asia
August 28, 2007	128	Total	Pacific, western North America
February 21, 2008	133	Total	North America
August 16, 2008	138	Partial	Europe, Africa
February 9, 2009	143	Penumbral	Western North America, Asia, Australia
July 7, 2009	110	Penumbral	Shading too light to be detectable
August 6, 2009	148	Penumbral	Eastern North America, Europe, Africa
December 31, 2009	115	Partial	Europe, Asia, Africa
June 26, 2010	120	Partial	North America, Australia, Pacific
December 21, 2010	125	Total	North America, Pacific

June 15, 2011	130	Total	Asia, Africa, Indian Ocean
December 10, 2011	135	Total	Asia, Australia, western North America
June 4, 2012	140	Partial	Australia, Pacific, western North America
November 28, 2012	145	Penumbral	Asia, Australia, western North America
April 25, 2013	112	Partial	Africa, Indian Ocean, Asia
May 25, 2013	150	Penumbral	Shading too light to be detectable
October 18, 2013	117	Penumbral	North America, Africa, Europe, Asia
April 15, 2014	122	Total	Pacific, North America
October 8, 2014	127	Total	Pacific, western North America
April 4, 2015	132	Total	Pacific, far west North America
September 28, 2015	137	Total	Eastern North America, Europe, Africa
March 23, 2016	142	Penumbral	Pacific western North America
February 11, 2017	114	Penumbral	North America, Europe, Asia, Africa
August 7, 2017	119	Partial	Africa, Asia, Australia
January 31, 2018	124	Total	Asia, Australia, far west North America
July 27, 2018	129	Total	Europe, Asia, Africa, Australia
January 21, 2019	134	Total	United States, South America
Universal Date:	**Saros**	**Type**	**Description**

July 16, 2019	139	Partial	South America, Europe, Asia, Africa
January 10, 2020	144	Penumbral	Europe, Asia, Africa, Australia
June 5, 2020	111	Penumbral	Shading probably too light to be visible
July 5, 2020	149	Penumbral	Shading probably too light to be visible
November 30, 2020	116	Penumbral	North America

Appendix D:
A Canon of Eclipses I Have Seen

H ere is a list of the 54 times that I've been touched by the wandering shadows of Earth and Moon. This personal canon of eclipses begins with an almost clouded out partial solar eclipse on October 2, 1959. On March 12, 1960, Mother promised me a lunar eclipse, but since I do not recall seeing it, that event might have clouded out also. (The fire eclipses that are lettered, not numbered, are events I planned to see but did not see.) However, according to this canon, I did see that eclipse-perhaps not in 1960, but when lunar saros 122 repeated itself two cycles later on April 4, 1996.

My first total solar eclipse, on July 20, 1963, was a part of saros 145, on July 20, 1963. I saw it again on August 11, 1999. When the cycles of the solar system converge again, I hope to enjoy the eclipse one more time, when it crosses the United States on August 21, 2017.[1]

---

1. The term Canon was first used in 1887 to describe a list of eclipses by Theodor von Oppolzer, whose Canon of Eclipses listed all solar eclipses, and all but the penumbral lunar eclipses from 1208 BC to 2161.

1. October 2, 1959. Session No. 1S. Solar Saros 143. This partial solar eclipse was eclipsed by clouds until its final fifteen minutes. A beautiful sunrise view of a partially eclipsed Sun.

    A. March 12/13, 1960. Lunar Saros 122. My mother promised me a view of this eclipse, but I do not recall seeing it—the sky must have been cloudy.

2. August 25/26, 1961. Session *7EM. Lunar Saros 137. A partial eclipse of the Moon, but since 99.2% of the Moon was covered by the shadow of the Earth at its maximum, this eclipse was as close to total as you can get without actually being total. Sky was partially cloudy. I watched the first part of the eclipse from the Observatory of the Montreal Center of the Royal Astronomical Society of Canada, which coincidentally happened to be located a hundred yards to the west of McGill University's Molson stadium, in whose stands my parents were, at the same time, watching a Montreal Alouettes football game. The game ended before the eclipse did, so we left and watched the rest of it from home.

3. July 20, 1963. *338S. Solar Saros 145. Total eclipse of the Sun, at Lake William, Quebec. See chapter 2.

4. December 30, 1963. *409M. Lunar Saros 124. Total eclipse of the Moon. One of the darkest eclipses ever, thanks to an volcanic eruption which filled Earth's atmosphere with sulfur. At totality, the Moon was completely invisible to me. According to Constantine Papacosmas, who saw the eclipse from a dark site, the

full Moon, which is usually as bright as magnitude minus 12, was as faint as a fifth magnitude star. From our Montreal home, the weather was very cold that night.

5. June 24, 1964. *519E. Lunar Saros 129. Total eclipse of the Moon. Despite a clear morning, clouds and thundershowers moved in by early evening, completely washing out our chances of seeing the total part of the eclipse. The clouds did clear in time for our group did see the last part of the partial phase.

6. December 18/19, 1964. *720EM. Lunar Saros 134. Total eclipse of the Moon. The last of an unusual series of three total lunar eclipses visible from a single site. Sky was clear. Our group, the Amateur Astronomers Association, enjoyed the eclipse and made a tape recording of the event. When the shadow passed over the sea of Crisis, a call of Mare Crisium sounded very much like Merry Christmas, and we patriotically finished the event with a rousing rendition of the Canadian national anthem.

7. April 12/13, 1968. 2045EM. Lunar Saros 131. Total eclipse of the Moon. A beautifully clear and warm Passover evening; the eclipse waited until after Seder ended. We went to a couple of sites—at one, it appeared that half of the people of Montreal were out watching!

8. March 7, 1970. 2275S2. Solar Saros 139. Total eclipse of the Sun. Although our group was under the shadow of the Moon, clouds obscured our view of the totally eclipsed Sun. However, I have never seen a

darker total eclipse. The thick layer of stratus clouds amplified the effect of the Moon's shadow, which swooped in from the southwest and plunged us into a late twilight. After totality ended, we could see the shadow racing away over the distant clouds.

9. August 16, 1970. *2328EM. Lunar Saros 110. Partial eclipse of the Moon. This event coincided with another one: The official opening of the Camp Minnowbrook summer Olympics! My preparations for observing the eclipse were hindered somewhat by my also having to help organize children for the games.

10. February 10, 1971. Lunar Saros 123. Total eclipse of the Moon. "Just before 2 A.M.," I wrote in my Journal, "the Earth's main shadow attacked, and in a slow but steady advance darkened the Moon to a beautiful coppery red. And before the Moon could thrust away this bloody cloak, the Earth sent clouds to cover the sky, thereby preventing anyone from witnessing the Moon's ultimate victory in its brief battle with the Earth."

11. January 30, 1972. 1972–12 Lunar Saros 133. A total eclipse of the Moon, seen from the roof of the 14-story Tower residence at Acadia University. The eclipse was accompanied by astronomical tides. After the eclipse ended we drove around the desolate, frozen wasteland that was the Minas Basin and nearby Grand Pre at low tide. Tall frozen icebergs seemed to stand on a desert of white, and to the west of the eerie scene, the Moon, almost emerged from the partial phase of the eclipse, was sinking into the west.

12. July 10, 1972. RVH-1. Solar Saros 126. A partial eclipse of the Sun. I had hoped that this would be my third total eclipse, but since I was ill I watched the eclipse from hospital grounds. This eclipse was, in fact, the one glorious moment of an otherwise unhappy summer. It didn't start that way—dense clouds and rain prevented any view of the eclipse at first, but at 4:55 PM, just at the moment of maximum eclipse, with the Sun more than 80% covered, the clouds broke and I was able to see the crescent Sun and photograph it.

B. July 25, 1972. Lunar Saros 138. Partial eclipse of the Moon. Clouded out.

C. January 18, 1973. Lunar Saros 143. Penumbral eclipse of the Moon. Clouded out, but saw full Moon later.

D. Friday, December 13, 1974. Solar Saros 151. A partial eclipse of the Sun. This was to be a great opportunity for public viewing; Constantine Papacosmas and I set up an 8-inch telescope in front of the Arts Building at McGill University, the single most heavily traveled point of the whole institution. We explained telescopes and eclipses to many people, but heavy clouds obscured the entire event. This last eclipse in this particular saros series, incidentally, took place as a small partial eclipse near the pole on December 2, 1956. The following one was an eclipse of greater magnitude on December 23, 1992. It will not be until May 16, 2227, that this saros will show its first total eclipse.

13. May 24/25, 1975. Lunar Saros 130. A total eclipse of the Moon. A contrast to the last eclipse, this one was observed from Jarnac Pond, Quebec, under a completely clear sky. SKYWARD.

14. November 18, 1975. Lunar Saros 135. A total eclipse of the Moon, seen through a hazy and partly cloudy sky from Montreal, Canada.

15. April 3/4, 1977. *3234E2. Lunar Saros 112. A Passover partial eclipse of the Moon.

16. October 12, 1977. *3451S. Solar Saros 143. A partial eclipse of the Sun, viewed from my home in Amherstview, Ontario, and from Wendee's home in Las Cruces, New Mexico. (12 pics)

17. February 26, 1979. ***3861S. Solar Saros 120. The last total eclipse of the Sun to touch the North American mainland in the 20th century crossed its way from Washington State, and into Manitoba. I observed this eclipse from Lundar, Manitoba. "During your lifetime," I wrote, "sometime while you can still walk and breathe, you must try to observe a total eclipse of the Sun. I have yet to see a spectacle that rips to the core of my being more thoroughly than such an event. As long as I live I shall never forget my feelings as I was gripped by the shadow of the Moon during the 1979 eclipse."

18. September 5, 1979. *4050AN3. Lunar Saros 137. A total eclipse of the Moon, seen with astronomer Gerald Cecil a few days after I moved to Tucson.

19. August 25/26, 1980. Session *4782E. Lunar Saros

147. Penumbral Lunar Eclipse. Definite but very slight shading. Watched this one while nursing the flu, lying on my back on a couch by a window.

20. January 19/20, 1981. *5098M. Lunar Saros 114. Penumbral Eclipse of the Moon. Observed and photographed from my home near Tucson.

21. July 16/17, 1981. Sessions *5302 EM3, * 5303M4, 5304MP5, 5305MP6, and 5306M7. Lunar Saros 119. Partial eclipse of the Moon observed, with friend Carl Jorgensen, from several sites around Montreal.

22. July 5, 1982. Sessions *5894AN2. Lunar Saros 129. A total eclipse of the Moon, which I saw after the afternoon's rain clouds grew, then dissipated. This was also the eclipse where astronomer Brent Archinal, first met his future wife JoAnne at an eclipse viewing party in Columbus, Ohio. Archinal observed it from the roof of the Physics building with friends from the Ohio State University astronomy club. The dome grew crowded as students from astronomy classes joined them. Late that evening, Archinal walked two of the women back to their dorm. "One of the women was interested in talking to me," Archinal remembers. "I found out her name was JoAnne"

23. December 30, 1982. Session *6117EM Lunar Saros 134. This total eclipse of the Moon was visible only in its partial phase from Tucson, since heavy clouds obscured the central part. Observed with comet discoverer Rolf Meier.

24. June 24, 1983. Session **6347AN. Lunar Saros 139.

Partial Eclipse of the Moon. Transits of two of Jupiter's moons were taking place simultaneously, with shadows of both visible on the face of Jupiter. So we saw three shadows that night—two of Jupiter's moons, and one of the Earth.

25. December 19/20, 1983. Session 6468E. Lunar Saros 144. Penumbral eclipse of the Moon. Sky cleared enough to get good views just after maximum eclipse. Structural detail seen on the Moon's dark edge.

E. May 15, 1984. Lunar Saros 111. No record of me seeing this penumbral eclipse. Was it cloudy?

26. May 30, 1984. *6594M. Solar Saros 137. Annular eclipse of the Sun. I saw this event from New Orleans, La. The shadow swooped out of the sky and left us, all in a few seconds. The Moon and the Sun were virtually identical in apparent size for this unusual eclipse. "Time stopped," I wrote, "then started again as the annular phase ended. Did it last five seconds, or a week?" Observing from Ias Cruces, Wendee wrote: "Today was exciting. There was a partial solar eclipse this A.M. The sky got dusk-like and it got breezy."

27. April 24, 1986. Session *7098M2. Lunar Saros 131. Moderately bright; Danjon lumonisity scale 2½ to 3. Some stratocumulus clouds but generally a fine view. Dawn began about 15 minutes after totality began.

28. October 3, 1986. Session *7249S-B. Solar Saros 124. Partial eclipse of the Sun. This was an eclipse with a second of totality far to the northeast of my site at the southwestern edge of New Mexico.

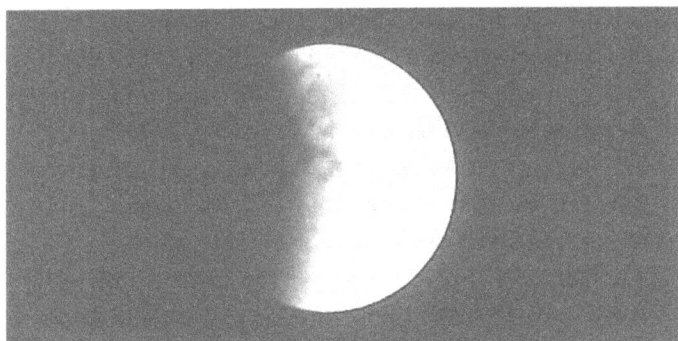

The progress of the July 1982 eclipse from ten minutes after totality to the late partial phase is shown in these three views. Photographs by Brent Archinal.

29. April 14, 1987. Session *7397E. Lunar saros 141. Penumbral lunar eclipse, again on Passover. Eclipsed Moon beautiful rising over the mountains to the east. Used an antique Ramsden telescope, some two centuries old, to view this eclipse.

30. October 6/7, 1987. Session 7533E. Lunar Saros 146. Penumbral lunar eclipse. Sharp darkening noticed. Clear shapes of mountains seen on lunar edge, as with eclipse no. 29 and other penumbral eclipses. Discovered a comet three nights later. Steve and Donna O'Meara were married on the day of this eclipse. Famed Astronomers and vol, they were married on an erupting volcano.

31. August 26/27, 1988. Session *7786M Saros 118. San Francisco, California. I set the alarm for 4 AM but there were dense low clouds. I decided to go back to sleep, but 10 minutes later I got up again and went outside, just in case. Now there were breaks in the rapidly moving cloud bank, and soon the 1/3 eclipsed Moon came into view! It was very nice, the eclipsed Moon in one direction, San Francisco skyline in another, and a fog horn sounding every 20 seconds.

32. February 20, 1989. *7945M. Lunar saros 123. Interesting effect just before sunrise—Earth shadow seen near Earth, and partially covering Moon too!

33. March 7, 1989. Session *7956S. Solar Saros 149. Partial eclipse of the Sun, with a large sunspot group on the Sun at the same time. (Observed with my two cats.)

34. August 16/17, 1989. Session **8054EM. Lunar Saros 128. Nova-searcher Peter Collins and I left Tucson under solid clouds. Forecasts showed that heading west would give us the best chance of clear sky. We drove through a massive lightning storm. Immediately after the rain stopped, sky cleared from west, and we saw the eclipsed Moon just after third contact. 85% of the Moon was still covered by the Earth's shadow. Setting up on the side of a deserted road, we observed a marvelous sky with lightning and the eclipsed Moon.

35. July 21/22, 1990. Session **8318SE. Solar Saros 126. Partial eclipse of the Sun. Eclipse began at 8:05:30 P.M. with the sun already partly below the horizon—A "marvelous tension" as the Sun started to set and we wondered if we had somehow miscalculated and would miss it. With the "Moon illusion" effect, the Sun appeared to be much larger than normal since it was near the horizon, the Moon appeared to cut quite a way into the Sun, as seen in the photograph, in the three minutes we had to see this eclipse.

36. July 11, 1991. Session ***8597S. Solar Saros 136. "The Big One"—with a totality of almost 7 minutes, this solar eclipse, near noon in La Paz, Mexico, was a marvellous event. The sky at totality was not as dark as expected, due to atmospheric effects from the recently erupted Pinatubo Volcano in the Philippines.

37. January 4, 1992. Session 8724S. Solar Saros 141. Annular eclipse of the Sun seen from, the west side of

Palomar Mountain, with Gene and Carolyn Shoemaker, Jean Mueller, Lonny Baker and Todd Hansen, and Tim and Carol Hunter. The sky was very cloudy until the moment of maximum eclipse, when the Sun appeared, for most of the several minutes of annular eclipse.

38. June 15, 1992. Session *8843EM. Lunar Saros 120. Unusally dark partial eclipse of the Moon, due to atmospheric effect from Pinatubo.

39. December 9, 1992. Session *8908E. Lunar Saros 125. Total eclipse of the Moon. Observed the eclipse just after moonrise with Clyde Tombaugh, discoverer of planet Pluto, and his wife Patsy. Wendee tried to see it from her home that night.

40. June 4, 1993. Session *9005M2. Lunar Saros 130. Predawn total eclipse of the Moon. Timed contacts of the shadow on Tycho, Plato, Mare Serenitatis. A really lovely eclipse.

41. May 21, 1993. Session *8987ANS. Solar Saros 118. Partial eclipse of the Sun seen through a bank of fog from Palomar Observatory.

42. November 29, 1993. Session *9094EM. Lunar Saros 135. Total Eclipse of the Moon. 1.5 on Danjon scale, meaning a quite dark eclipse.

43. May 10, 1994. Solar Saros 128. Session 9214S. Annular eclipse of the Sun viewed from Las Cruces, New Mexico, with Clyde and Patsy Tombaugh, and Brad Smith, imaging team leader of the Voyager spacecraft. Nearby, Wendee Wallach was leading an observing ses-

sion with children from Sierra Middle School, during their physical education class.

44. May 24/25, 1994. *9224EM. Lunar Saros 140. Very slight partial eclipse of the Moon, just a small amount of Moon covered by the shadow of the Earth. Observed with variable star observers John Griese and Charles Scovil. Very nice through small telescope.

45. November 18, 1994. Session *9296M. Lunar saros 145. A penumbral eclipse of the Moon, observed with astronomers Peter and Dianne Jedicke. Very slight penumbral shading on north side of Moon.

46. April 4, 1996. Session **9673E. Lunar Saros 122. Another Passover total eclipse of the Moon seen from Montreal. Moon was a medium-bright red; brighter lower part, much darker at top. Saw Comet Hyakutake as well.

47. September 27, 1996. Session *9813E. Lunar Saros 127. Total eclipse of the Moon. I observed this eclipse through dense clouds after a lecture at Ball State University in Indiana. The audience and I watched as the Moon approached total eclipse. Meanwhile, Wendee could see the entire eclipse under a clear sky from our home in Vail, Arizona. "Once it was about 75% eclipsed, it looked like a carrot cake cupcake with white icing," she wrote. "Once it was total, the Moon looked like a hige piece of amber hanging in the sky.... Once the Moon began brightening [after totality,] the copper glow turned into silver/white."

48. March 23/24, 1997. Session ***10063SEM2. Lunar Saros 132. Partial eclipse of the Moon. This was the prime event at our wedding reception, held outdoors at our home. The sky was beautiful. The eclipse was spectacular, and we also saw Comet Hale-Bopp.

49. February 26, 1998. Session *10,401MS. Solar Saros 130. Total Eclipse of the Sun seen from the Dawn Princess near Aruba. Sky was a dark, crisp blue at totality, with several planets visible. Corona and prominences were spectacular.

50. March 12/13, 1998. Session *10,413EM. Lunar Saros 142. Penumbral Eclipse of the Moon. Not really detectable with naked eye, but dark and beautiful with telescope. Edge opposite shadow was bright by contrast with rest of Moon. Seen through clouds.

51. September 5/6, 1998. Session *10,642M2. Lunar Saros 147. Penumbral Lunar Eclipse. Lunar rays pronounced during this ecliose, as they are with most penumbral ecliopses.

52. August 11, 1999. Session ***11177SANS. Solar Saros 145. Total Eclipse of the Sun, as described in this book.

53. January 20/21, 2000. Session *11434EM2. Lunar Saros 124. Total Eclipse of the Moon.

54. July 15/16, 2000. Lunar Saros 129. This is the third time I've seen this Total Eclipse of the Moon, but it was visible just as a partial from Arizona, where Wendee and I saw it with Stephen James O'Meara. Donna O'Meara, Steve's wife, observed the total phase

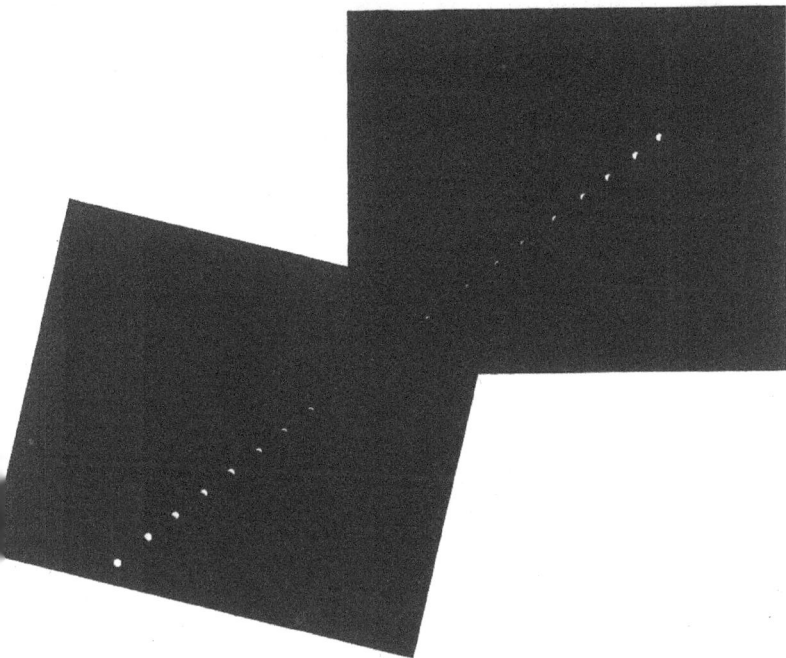

Lunar Eclipse of March 23, 1997. Multiple exposure by Keith Schreiber.

from their Hawaii home. "The moon looked like it was dipped in Hershey's milk chocolate, the reddish-brown kind. Compare with Wendee's description of eclipse No. 47.

Appendix E:
Eclipse Song

During the ship's variety show on the last day of our cruise, Gertrude Bishop sang these words in honor of Roy. Tune: *Tara's Theme (My Own True Love) from Gone with the Wind.*

> *I see the moonlight*
> *I see the starshine*
> *The Northern Lights*
> *Upon the sea*
>
> *But when the Sun*
> *Is kissed by Moon*
> *It fills the soul*
> *With wondrous joy*
>
> *I roamed the Earth*
> *in search of this*
> *I knew I'd know you, know you*
> *By the light*

And by the light
Eclipsed at sea
A sacred moment
Will ever be

The light of Moon
Married to Sun
They danced at dawn, dawn
On the sea

And by this gift
Of shining light
We're each a part of
The Universe.

INDEX

INDEX

INDEX

INDEX

INDEX